圜丘正立面
Front elevation of Circular Mound Altar

皇穹宇组群立面渲染图
Rendering of Elevation of Complex of Imperial Vault of Heaven

成贞门正立面
Front elevation of Chengzhen Gate

皇穹宇正立面
Front elevation of Imperial Vault of Heaven

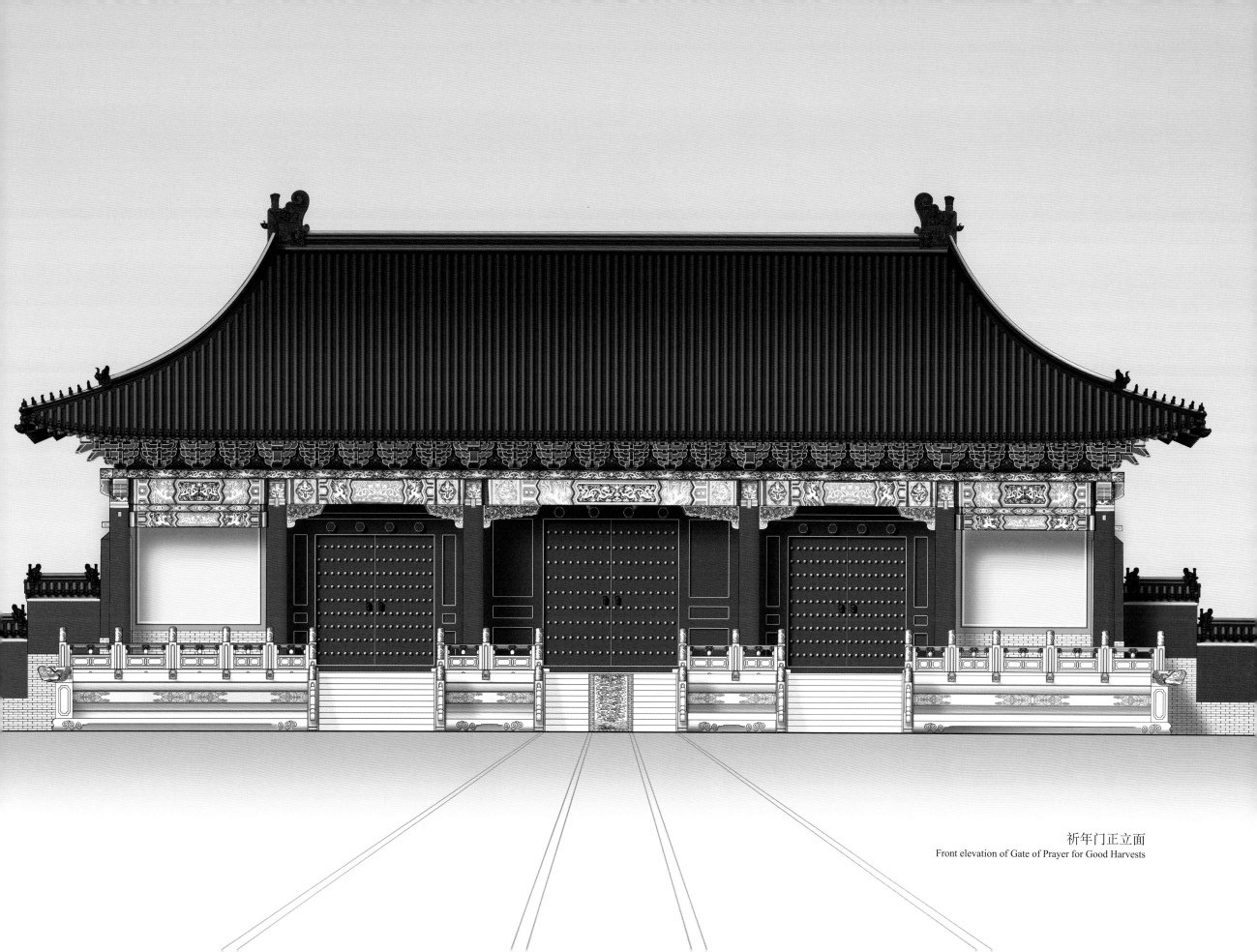

祈年门正立面
Front elevation of Gate of Prayer for Good Harvests

祈年殿祈谷坛正立面
Front elevation of Hall of Prayer for Good Harvests and Altar of Prayer for Good Harvests

皇乾殿正立面
Front elevation of Hall of Imperial Zenith

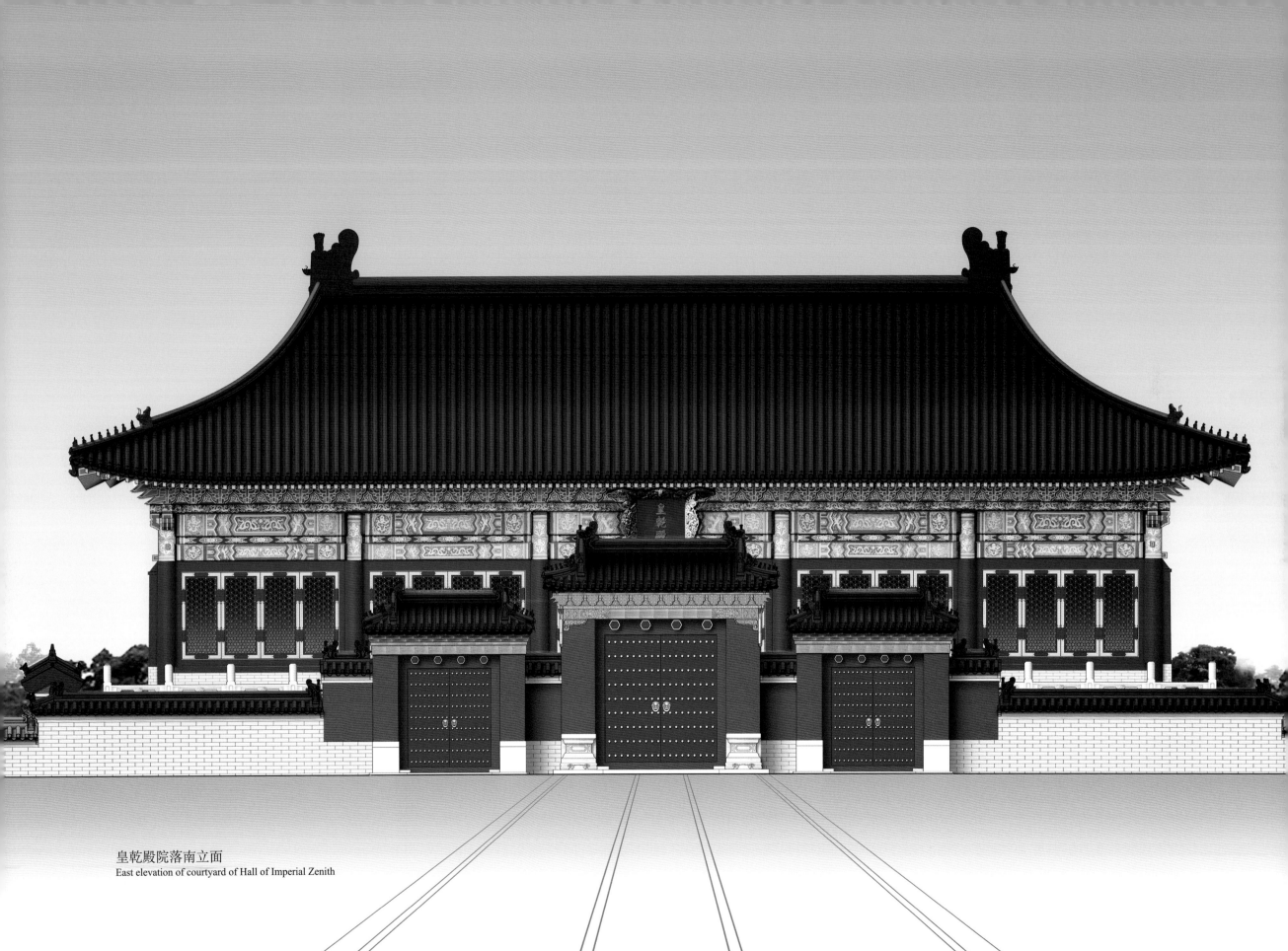

皇乾殿院落南立面
East elevation of courtyard of Hall of Imperial Zenith

祈年殿组群立面渲染图
Rendering of Elevation of Complex Hall of Prayer for Good Harvests

国家出版基金项目

『十二五』国家重点图书出版规划项目

中国古建筑测绘大系·坛庙建筑

天坛

天津大学建筑学院 北京市天坛公园管理处 编写

王其亨 主编

曹鹏 编著

中国建筑工业出版社

『十二五』国家重点图书出版规划项目

Traditional Chinese Architecture Surveying and
Mapping Series:
Temples Architecture

The Temple of Heaven

Compiled by School of Architecture, Tianjin University &
The Administration of the Temple of Heaven
Chief Edited by WANG Qiheng, Edited by CAO Peng

China Architecture & Building Press

Contents

Location: East side of Yongding-Gate Inner Street, Dongcheng District, Beijing

Construction Date: The 18th year of Yongle reign, the Ming Dynasty(1420)

Area: About 273 hectares

Responsible Department: Administration of the Temple of Heaven Park

Survey Department: School of Architecture, Tianjin University

Survey Time: 1986-2017

地　　址　北京市东城区永定门内大街东侧

始建年代　1420 年（明永乐十八年）

占地面积　约 273 公顷

主管单位　北京市天坛公园管理处

测绘单位　天津大学建筑学院

测绘时间　1986—2017 年

Preface

The Temple of Heaven (*Tiantan*) is located at the eastern side of Yongding-Gate Inner Street, opposite from the Temple for the Divine Cultivator (*Xiannongtan*) in Beijing. It covers an area of 273 hectares and is surrounded by two cordons of walls, both of which are semi-circular in the north, representing heaven, and rectangular in the south, representing earth. The two layers of walls divide the Temple of Heaven into an inner temple space and an outer temple space. In the inner temple space, buildings assembled include the three main complexes—the Circular Mound Altar (*Yuanqiu*), the Hall of Prayer for Good Harvests (*Qiniandian*), the Palace of Abstinence (*Zhaigong*) and accessory buildings such as the Divine Kitchen (*Shenchu*), the Divine Storeroom (Shen ku) and the Pavilion of Immolation (*Zaishengting*). The outer temple space consists of the Divine Music Administration (*Shenyueshu*), which serves sacrificial music to royal ceremonies, and the Building for Sacrificial Livestock (*Xishengsuo*), which serves offerings. Apart from the buildings, there are various historic trees in the Temple of Heaven, including pines, cypresses, Chinese pagoda trees and elms. These trees are either planted in line for ritual uses or planted freely on open ground as woods.

Constructed in the eighteenth year of the reign of the Yongle Emperor (1420) in the early Ming Dynasty, and well maintained by the following sovereigns, the Temple of Heaven along with its magnificent heaven-sacrificing buildings and dignified nature-worshipping open spaces has become the largest, the best preserved and the most cultural-connotation abundant ancient sacrificial architecture in China, and even in the world. It distinctly reflects the relationship between man and heaven—the human world and God's world—which stands at the heart of Chinese philosophy, and genuinely demonstrates the profound Chinese civilization. The Temple of Heaven was designated as one of China's foremost-protected cultural heritage sites by the State Council in 1961, and was listed as a UNESCO World Cultural Heritage Site in 1998 (fig.1).

导　言

天坛位于北京城正阳门以南，永定门内大街东侧，与先农坛隔街相望，占地面积 273 公顷。坛域由北圆南方的两道坛墙嵌套围合，形成内坛与外坛，内坛仅有圜丘、祈年殿、斋宫三大组群，及神厨、神库、宰牲亭等附属建筑；外坛设置了神乐署与牺牲所组群，为整个皇家祭祀活动提供乐舞与牺牲；其余广阔空间大量栽植松、柏、槐、榆等仪树与海树。

尊崇庄重、典雅瑰丽的天坛祭天建筑与苍然肃穆的郊天场所，始建于明永乐十八年（1420 年），经明清两代持续经营，凝聚了古代中国『参天地赞化育』的文化传统与生态智慧，体现了华夏文明的深厚积淀，堪称中国乃至世界现存规模最大、保存最完整、文化内涵最丰富的古代祭天建筑组群，1961 年被国务院公布为第一批全国重点保护文物单位，1998 年被联合国教科文组织（UNESCO）认定为世界文化遗产（图一）。

图一　1943年北京城航拍图

Fig.1　Aerial View of Beijing in 1943

I. Hongwu Emperor's Reign, the Ming Dynasty: Before the Temple of Heaven

The concepts of worshipping nature in suburbs went through a great change from sacrificing heaven and earth respectively to sacrificing them together in the reign of the Hongwu Emperor, the Ming Dynasty. The reformation in sacrificing concepts led to the transformation of the related architecture during this period, among which the Circular Mound Altar (*Yuanqiu*) in the southern capital of the Ming Empire (today's Nanjing), the Circular Mound Altar in the middle capital of the empire (today's Fengyang) and the Hall of Great Sacrifices (*Dasitan*) in the southern capital of the empire were the most crucial ones and had significant influences on the current Temple of Heaven (*Tiantan*) in Beijing.

1. The Circular Mound Altar in the Southern Capital

In the twenty-sixth year of the reign of the last emperor of the Yuan Dynasty (1366), ZHU Yuanzhang, the future emperor of the Ming Dynasty, almost conquered the land of China and therefore started thinking about his future empire. To start with, he decided to appoint Jinling (modern Nanjing) as his future southern capital and renamed it Yingtian. Then he ordered to extend the boundaries of Yingtian and to divine appropriate places for his imperial buildings. The next year (1367), he put forwards an edict to start the constructions of the imperial sacrificial altars and temples for nature, for harvests and for ancestors in Yingtian. Soon after the edict was given, in the eighth month of that year, the Circular Mound Altar (*Yuanqiu*) for heaven and the Square Mound Altar (*Fangqiu*) for earth were completed. Based on the researches on the ancient rituals for worshipping nature and the Chinese philosophy of Yin and Yang, the Circular Mound Altar was built at the south of the Mount Zhong in southern suburb while the Square Mound Altar was built at the north of the Mount Zhong in northern suburb. Five months later (1368), ZHU Yuanzhang offered a great rite to heaven and earth and proclaimed himself the Hongwu Emperor in the Circular Mound Altar, showing his legitimacy and authority as the emperor of the new Ming Dynasty, the real Son of Heaven (fig.2).

Located at the southeast of Yingtian, the Circular Mound Altar comprised an altar and two layers of walls enclosing the altar. Based on the rituals from the Han Dynasty, the central altar of the Circular Mound Altar in Yingtian was designed into an empty circular platform on two levels of glazed terraces, while the two layers of walls were designated

一、明洪武朝的郊天建筑

明洪武一朝的郊祀制度经历了从初制『天地分祀』到定制『天地合祀』的重大变化，进而由祀典引发了建筑的变化，其间营建的南京圜丘、中都圜丘和南京大祀坛对后世北京天地坛与天坛有着极为重要的影响。

1. 南京圜丘

元至正二十六年（1366 年），朱元璋初定天下，拟定都金陵，遂卜新宫，拓旧城。吴元年（1367 年）始建郊坛、社稷与宗庙。稽考郊祀之礼，顺合阴阳之义，在京师钟山之南营建圜丘，钟山之北营建方丘，分祀天地。同年八月，南郊圜丘与北郊方丘建成。洪武元年（1368 年）正月，朱元璋大祀天地于南郊圜丘，即皇帝位（图二）。

初建之圜丘，在正阳门外东南，钟山之阳。圜丘坛仿汉代制度，为圆坛两层，琉璃材质。坛外有圆、方两重壝墙，四面皆设棂星门。壝墙外北有天库，东置神厨、神库、宰牲房与天池，坛壝以南的横甬路东西各有牌楼一座。洪武二年（1369 年），先是增建天下神祇坛于圜丘之东，在有事

图三 《圜丘之图》（引自：四库本《大明集礼》）

圜丘之圖

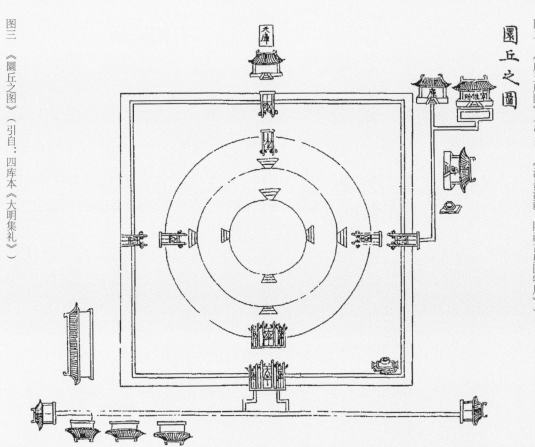

图二 《应天府图》（引自：《河岳海疆·院藏古舆图特展》）

Fig.2 *Map of Yingtian City* (Source: *Mapping the Imperial Realm: an Exhibition of Historical Maps*)
Fig.3 *Map of the Complex of Circular Mound Altar in Yingtian* (Source: *Collected Rites of the Ming Dynasty in Imperial Collection of Four*)

into a circular inner wall and a rectangle outer wall, symbolizing heaven and earth respectively. There was a Lingxing Gate in each side of the inner and outer walls serving as an entrance. Outside the Circular Mound Altar, there was the Storeroom of Heaven (*Tianku*) at the north; the Divine Kitchen (*Shenchu*), the Divine Storeroom (*Shenku*), the Pavilion of Immolation (*Zaishengting*) and the Pond of Heaven (*Tianchi*) at the east; two Chinese Decorated Archways (*Paifang*) on both sides of the transverse path at the south. In the second year of the reign of ZHU Yuanzhang, the Hongwu Emperor (1369), the Altar for All Gods (*Tianxiashenqitan*) was added as requested at the east of the Circular Mound Altar, which was used to give advance notices to all gods before the rites were held. Later on a hall with nine bays was built at the south of the Circular Mound Altar as an alternative place for rites when it rained. The next year (1370), the Palace of Abstinence (*Zhaigong*) was added at the west of the Circular Mound Altar and the Building for Sacrificial Livestock (*Xishengsuo*) at the south was rebuilt. In the sixth year of the reign of the Hongwu Emperor (1373), imitating the image of the *Jing* Bell—a divine bell—in the Song Dynasty, the Bell of Supreme Harmony (*Taihezhong*) was produced. To hang the bell, a Bell Tower (*Zhonglou*) was constructed at the northeast of the Palace of Abstinence. The completion of the Bell Tower marked the end of the construction of the Complex of Circular Mound Altar in Yingtian. At that point, the Circular Mound Altar served only as the altar for heaven (fig.3).

2. The Circular Mound Altar in the Middle Capital

With the dream of setting up two capitals as the ancient Zhou and Han dynasties did, on the first day of the eighth month of the first year of his reign (1368), ZHU Yuanzhang put forward an imperial edict to assign Yingtian (modern Nanjing) the southern capital of the Ming Dynasty and Bianliang (modern Kaifeng) the northern capital of the Ming Dynasty. The next day, XU Da, the general of the Ming, occupied *Dadu* (modern Beijing), the capital of Yuan, which led to ZHU Yuanzhang's uncontested control of the land of China. Under this condition, considering that Bianliang had not yet recovered from the war, ZHU Yuanzhang made an inspection tour through China in the following two months to find a city as an alternate for Bianliang. In the next year (1369), after a thorough comparison among different cities, ZHU Yuanzhang finally decided to replace Bianliang with his hometown city Linhao (today's Fengyang), and appointed it as the middle capital due to its geographical location. Then he started the constructions of city walls, moats, palaces and altars in the middle capital. In the fourth month of the eighth year of the reign of the Hongwu Emperor (1375), all constructions were called stop for the unanticipated

于圜丘之前预告百神之用，后在圜丘以南建殿九间，作为遇风雨时的望祭之所。翌年，建斋宫于圜丘之西并改建南郊牺牲房。洪武六年（1373 年），仿宋景钟铸成太和钟，即在圜丘斋宫东北建钟楼悬挂此钟。至是，以天地分祀为制度的南京圜丘建设基本完成（图三）。

2. 中都圜丘

洪武元年（1368 年）八月初一，朱元璋下诏以金陵为南京，大梁为北京。次日，大将军徐达率军攻取了元大都。在这种国家南北一统的态势下，其后的两个月，朱元璋先是巡幸北京，后又迁徙北平军民于汴梁，其欲于「天下之中」——宋之旧京立都之意尤为明显。然而，次年（1369 年）九月，朱元璋却以「民未苏息」为由，诏以家乡临濠（今凤阳）为中都，并开始建置城池、宫阙、坛庙。至洪武八年（1375 年）四月，以「劳费」为由，罢中都役作。虽然当时中都城池、宫阙和坛庙大多尚未完工或并未建设，但罢建之前数日，朱元璋刚刚在驻中都时祭告天地于圜丘，或可说明作为国家祭祀场所之首的圜丘业已建成。

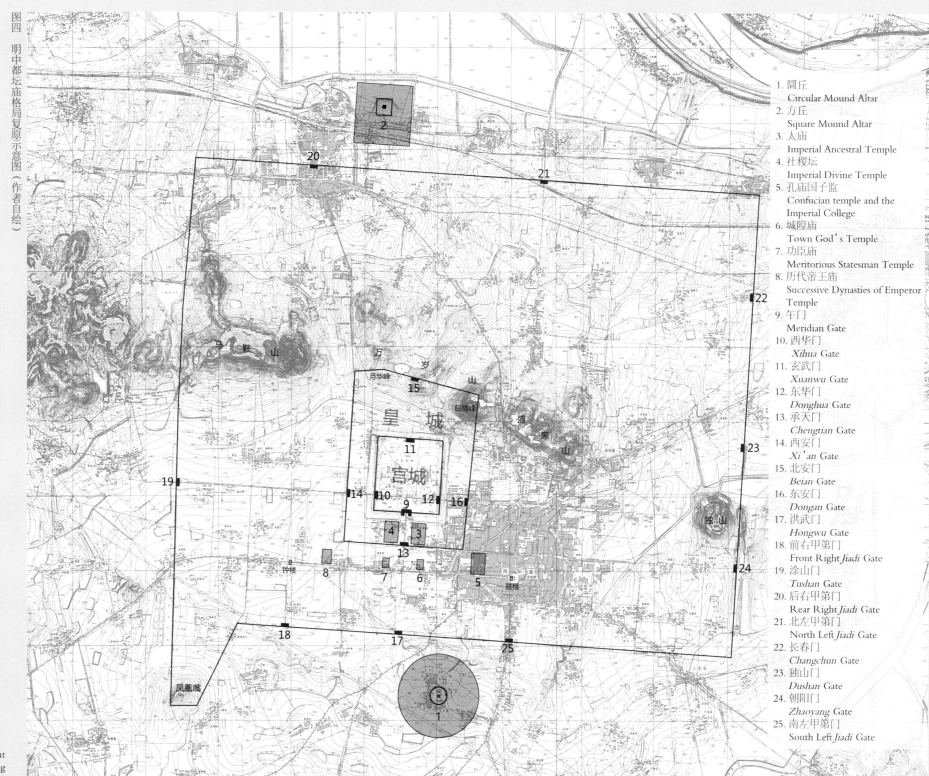

图四　明中都坛庙格局复原示意图（作者自绘）

1. 圜丘
   Circular Mound Altar
2. 方丘
   Square Mound Altar
3. 太庙
   Imperial Ancestral Temple
4. 社稷坛
   Imperial Divine Temple
5. 孔庙国子监
   Confucian temple and the
   Imperial College
6. 城隍庙
   Town God's Temple
7. 功臣庙
   Meritorious Statesman Temple
8. 历代帝王庙
   Successive Dynasties of Emperor
   Temple
9. 午门
   Meridian Gate
10. 西华门
    Xihua Gate
11. 玄武门
    Xuanwu Gate
12. 东华门
    Donghua Gate
13. 承天门
    Chengtian Gate
14. 西安门
    Xi'an Gate
15. 北安门
    Beian Gate
16. 东安门
    Dongan Gate
17. 洪武门
    Hongwu Gate
18. 前右甲第门
    Front Right Jiadi Gate
19. 涂山门
    Tushan Gate
20. 后右甲第门
    Rear Right Jiadi Gate
21. 北左甲第门
    North Left Jiadi Gate
22. 长春门
    Changchun Gate
23. 独山门
    Dushan Gate
24. 朝阳门
    Zhaoyang Gate
25. 南左甲第门
    South Left Jiadi Gate

Fig.4 Restored Map of the Architectural Layout
of Altars and Temples in Linhao, the Ming
Dynasty (Drawn by author)

huge costs. Though most of the buildings were left unfinished, the Circular Mound Altar (*Yuanqiu*) in Linhao was fully completed and even used once by ZHU Yuanzhang to sacrifice heaven and earth several days before the constructions in the middle capital was shut down.

According to records, the construction of the Circular Mound Altar in the middle capital, Linhao, was launched in the first month of the fourth year of the reign of the Hongwu Emperor (1371). The scale of the altar was slightly smaller than that of the Circular Mound Altar in the southern capital. Based on the relics remained at the southeast of Linhao, it can be inferred that there were three concentric circular structures in the Circular Mound Altar: the innermost one was a circular altar, which appeared to be a round mound now; the middle one was a circular ditch, on which there should be bridges in all four directions in the past; the outermost one was a circular wall about 6 *li* (3456 meters) in perimeter. Even though what stood between the ditch and the wall remained to be explored, the general concentric-circular layout of the Circular Mound Altar in Linhao had graphically shown the symbolic connection between heaven and circle (fig.4).

3. The Hall of Great Sacrifices in the Southern Capital

The failure of both Bianliang (modern Kaifeng) and Linhao (modern Fengyang) shattered ZHU Yuanzhang's dream of having more than one capital, and forced him to focus only on the remaining southern capital Yingtian (modern Nanjing). In the seventh month of the eighth year of the reign of the Hongwu Emperor (1375), the Imperial Ancestral Temple (*Taimiao*) was moved to the southeast of the imperial palace in Yingtian, and was transformed from a series of constructions in which each ancestor had his own sacrificial building into a single hall in which the ancestors were distributed into different sacrificial chambers. The change of the Imperial Ancestral Temple not only suggested the emperor's will to re-organize the city space of Yingtian, but also indicated the beginning of the transformation of sacrificing concepts from sacrificing different gods and spirits separately to sacrificing them together.

In the eight month of the tenth year of his reign (1377), ZHU Yuanzhang sent out an imperial edict to build a hall on the platform of the Circular Mound Altar (*Yuanqiu*) in Yingtian, to name it the Hall of Great Sacrifices (*Dasidian*), and to worship heaven and earth together there. In the tenth month of the next year (1378), the Complex of Hall of Great Sacrifices was completed. Since the hall was built for sacrificing heaven and earth

3. 南京大祀坛

罢建中都使朱元璋如周汉之制营建两京的蓝图成为泡影，他随后即把营建重点重新转回南京。

洪武八年（1375年）七月，首先改建了南京太庙于阙左，且由「都宫别殿」改为「同堂异室」，这不仅标志着南京大规模调整都城格局的开始，也标志着坛庙祀典制度由「分祀」转向「合祀」的开始。

从文献记载可知，中都圜丘始建于洪武四年（1371年）正月，其坛壝尺度较南京略有缩减。

从遗存情况来看，如今位于中都城洪武门外东南的圜丘坛垣，其总体格局形态尚能辨析，中部的圆形土岗是圜丘的内坛空间，即中间为圜丘坛所在；外以圆形环沟围绕，四面设桥与内坛相连；内坛之外尚有周回约六里的巨大圆形外坛。虽然圜丘遗址内的建筑情况尚不清晰，但通过遗址中内坛、环沟与外垣不断增大的圆形空间，已展现出在明初「天地分祀」制度下古代哲匠对「天圆」意象的完美演绎（图四）。

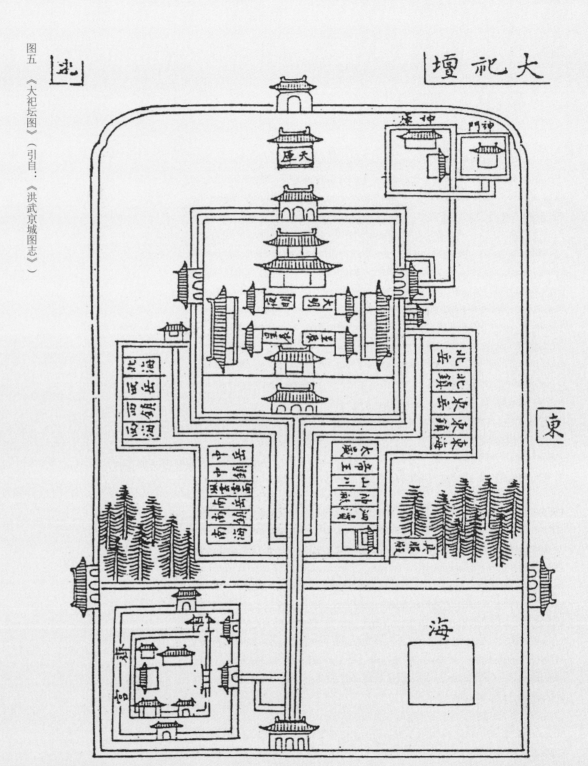

图五 《大祀坛图》（引自：《洪武京城图志》）

Fig.5 *Map of the Hall of Great Sacrifices* (Source: *Map of the Capital in Hongwu Emperor's reign*)

together, the date for offering rites in the hall was set in the first lunar month, early spring, so that the emperors could pray for blessings on agriculture activities. The new Complex of the Hall of Great Sacrifices could be divided into two parts, the inner temple and the outer temple. The inner temple contained of the Hall of Great Sacrifices, the Gate of Great Sacrifices (*Dasimen*), east and west wings and a consistent wall with four stone gates in it. The Storehouse of Heaven (*Tianku*) was outside the north gate of the inner temple. The outer temple consisted of the Divine Kitchen (*Shenchu*), the Divine Storeroom (*Shenku*) and the Pavilion of Immolation (*Zaishengting*) at the northeast; the Palace of Abstinence (*Zhai Palace*) at the southwest; and a consistent wall of 9 *li* and 30 *bu* (5232 meters) (fig.5). According to the literal and graphical records, it is clear that the Complex of Hall of Great Sacrifices is the ritual and architectural prototype of the current Temple of Heaven (*Tiantan*) in *Beijing*.

II. The Ming Dynasty: At the Time of the Temple of Heaven

The Temple of Heaven (*Tiantan*) in Beijing was initially built in the reign of the Yongle Emperor, the Ming Dynasty, in the southern suburb of the northern capital of the empire, imitating the Hall of Great Sacrifices (*Dasidian*) in Yingtian (modern Nanjing). By then, the temple was called the Temple of Heaven and Earth (*Tianditan*), because it offered rites to heaven and earth at the same time following the reformed rituals committed by ZHU Yuanzhang. Years later, ZHU Houcong—the man from a collateral branch of the royal family—became the Jiajing Emperor. To show his sovereignty of the Ming, he stipulated to offer separate sacrifices to heaven and earth once again. Under this condition, the Temple of Heaven and Earth was renamed the Temple of Heaven to show that the temple sacrificed the God of Heaven solely, meanwhile a new Circular Mound Altar (*Yuanqiu*) was added specially in the temple space to worship heaven. At a later time, according to the ancient records of the Palace of Enlightenment (*Mingtang*), the Hall of Great Sacrifices inside the temple was redesignated into the Hall of Great Offerings (*Daxiangdian*). Until then, the primary layout of the current Temple of Heaven was finally formed.

1. The Temple of Heaven and Earth in the Reign of the Yongle Emperor

In the seventeenth day of the sixth lunar month, the thirty-fifth year of the reign of the

二、明北京天地坛与天坛

北京天坛创建于明永乐朝，按太祖洪武定制合祀天地于南郊，名为『天地坛』，其建筑制度仿效南京大祀殿。至嘉靖朝，因旁系承继皇位而引发礼制变革，复行分祀天地的洪武初制，创建圜丘专祀天帝，天地坛亦更名为『天坛』，其后又稽古创建明堂，改建大祀殿为大享殿，奠定了北京天坛建筑的基本格局。

1. 明永乐朝天地坛

明洪武三十五年（1402年）六月十七日，朱棣在南京称帝。永乐元年（1403年）正月，大祀天地后，

里三十步（图五）。南京大祀殿的祀典制度与建筑形制成为后世永乐朝北京天地坛的蓝本。

庑，大祀殿，北门外为天库；内坛之外，东北为神厨库与宰牲亭等，西南为斋宫组群，坛垣周回九

事。此次改建的南京大祀殿组群亦分内坛与外垣，内坛以四座石门与周墙围绕，中有大祀门、东西

举合祀天地之典。翌年十月，大祀殿建成，并更定合祀之日为春首正月三阳交泰之时，以启稼穑之

洪武十年（1377年）八月，朱元璋下诏改建圜丘为大祀殿，即圜丘旧址为坛而以屋覆之，欲

图七　天地坛大祀殿组群总平面复原图（作者自绘）

图六　北京天地坛总平面复原图（作者自绘）

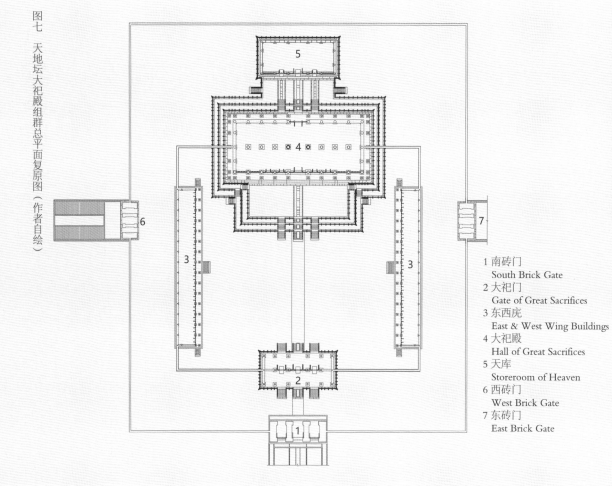

1 南砖门
　South Brick Gate
2 大祀门
　Gate of Great Sacrifices
3 东西庑
　East & West Wing Buildings
4 大祀殿
　Hall of Great Sacrifices
5 天库
　Storeroom of Heaven
6 西砖门
　West Brick Gate
7 东砖门
　East Brick Gate

1 大祀殿组群
　Complex Hall of Great Sacrifices
2 神厨库
　Divine Kitchen and Divine Storeroom
3 宰牲亭
　Pavilion of Immolation
4 具服殿
　Imperial Dressing Hall
5 斋宫组群
　Complex of Palace of Abstinence

6 祠祭署
　Sacrificial Administration
7 神乐观组群
　Complex of Divine Music Monastery
8 牺牲所组群
　Complex of Building for Sacrificial Livestock
9 南天门
　South Heavenly Gate
10 东天门
　East Heavenly Gate

11 北天门
　North Heavenly Gate
12 内西天门
　Inner West Heavenly Gate
13 外西天门
　Outer West Heavenly Gate

Fig.6 Restored Map of the Temple of Heaven and Earth in Shuntian (Drawn by author)
Fig.7 Restored Map of the Complex of Hall of Great Sacrifices in the Temple of Heaven and Earth (Drawn by author)

Hongwu Emperor (1402), ZHU Di, the son of ZHU Yuanzhang, proclaimed himself as the Yongle Emperor of the Ming Dynasty in *Yingtian* (modern Nanjing). In the first month of the next year (1403), after a grand sacrifice to heaven and earth, ZHU Di, the newly enthroned emperor, decided to follow the suggestions of his officials of rituals to assign a city as the other capital in addition to Yingtian, as his father had done with Linhao (today's Fengyang). After thoughtful comparisons, Beiping (today's Beijing), the garrison of ZHU Di when he was still a prince, was selected and assigned the northern capital, and was then renamed Shuntian. Shortly afterwards, with the ambition to accommodate the empire's center of gravity in Shuntian, ZHU Di took various measures to renovate the city. The officials were sent to redo the urban planning of Shuntian; to collect raw materials such as wood and bricks for building constructions; to restore old altars and temples; to build new imperial palaces and mausoleums according to the results of divinations; to dredge channels and to develop water transportation. By the end of the eighteenth year of the reign of ZHU Di, the Yongle Emperor (1420), the imperial palace in Shuntian, which is now known as the Forbidden City, had finally been completed. The next year (1421), ZHU Di moved to Shuntian along with his government and officials after a grand ceremony.

The Temple of Heaven and Earth (*Tianditan*) in Shuntian was built at the same time as the Forbidden City was constructed. The architectural design of the Temple of Heaven and Earth was an imitation of the Complex of Hall of Great Sacrifices (*Dasidian*) in Yingtian, with the same four main complexes—namely the Hall of Great Sacrifices, the Palace of Abstinence (*Zhaigong*), the Divine Music Monastery (*Shenyueguan*) and the Building for Sacrificial Livestock (*Xishengsuo*)—and other similar annexes, including the Divine Kitchen (*Shenchu*), the Divine Storeroom (*Shenku*), the Pavilion of Immolation (*Zaishengting*), the Storeroom of Imperial Carriages (*Luanjiaku*) and the Imperial Dressing Hall (*Jufudian*). Similar to the layout of the Complex of Hall of Great Sacrifices in Yingtian, the Temple of Heaven and Earth was divided into an inner temple and an outer temple by two layers of walls: the Hall of Great Sacrifices was inside the inner temple, while the three other main complexes belonged to the outer temple. However, in regard to the sittings, the scales and the ambient environments, the Temple of Heaven and Earth in Shuntian was much superior than the Hall of Great Sacrifices in Yingtian. To be specific, in terms of sittings, with the experience gained from the temples in Yingtian and Linhao, the Temple of Heaven and Earth was located at the east of the southern suburb of Shuntian city, opposite to the Temple of Mountains and Rivers (*Shanchuantan*) on the other side of the main axis of the city. The symmetric distribution of the two temples in Shuntian was substantially dignified and organized than the distribution in Yingtian where

朱棣在礼臣建议下，遵太祖中都之制，升运兴之地的北平为北京，形成『两京之制』。其后堪合皇城，采办砖木，疏浚河道，修治坛庙，整理城池，营建山陵，通漕济运，诏作西宫，鼎建新宫。至永乐十八年（1420 年）末，北京宫殿成。翌年初，朱棣隆重迁都北京。

北京天地坛虽然在祀典制度和建筑配置上『悉如南京』，形成了大祀殿、斋宫、神乐观、牺牲所四大组群以及神厨库、宰牲亭、具服殿等附属建筑，但在规划选址、坛垣尺度、外部空间等多个方面体现出『高敞壮丽过之』的设计超越。在规划选址上，北京天地坛吸取了南京、中都坛庙布局的经验，置于正阳门外南郊之东，与西侧的山川坛隔街相望，共同拱卫北京的城市中轴线，较之南京两坛皆规划于正阳门外之东更具仪式感；在坛垣尺度上，北京天地坛由南京大祀坛的周回『九里三十步』扩大到周回『十里』，且形成了内、外坛空间，以区分主体建筑大祀殿、斋宫与附属建筑牺牲所、神乐观（图六）；在外部空间上，大祀殿组群置于高台之上，南神门通向大祀门的神道逐渐抬升，更加增强了人神对话空间的崇高感（图七）。

明嘉靖九年至十年（1530—1531年）鼎建圜丘

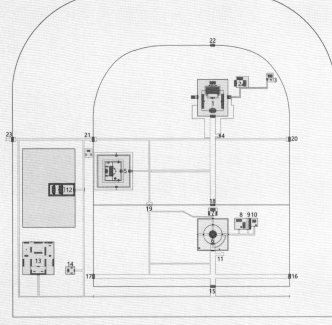

明嘉靖十一年（1532年）增设崇雩坛

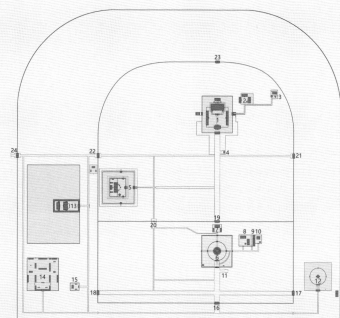

1 大祀殿　Hall of Great Sacrifices
2 神厨库　Divine Kitchen and Divine Storeroom
3 宰牲亭　Pavilion of Immolation
4 具服殿　Imperial Dressing Hall
5 斋宫　Palace of Abstinence
6 圜丘　Circular Mound Altar
7 泰神殿　Hall of Imperial Portraits and Tablets
8 神厨库　Divine Kitchen and Divine Storeroom
9 三库　Triple Storeroom
10 宰牲亭　Pavilion of Immolation
11 具服台　Imperial Dressing Platform
12 神乐观　Divine Music Monastery
13 牺牲所　Building for Sacrificial Livestock
14 祠祭署　Sacrificial Administration
15 昭亨门　Zhaoheng Gate
16 泰元门　Taiyuan Gate
17 广利门　Guangli Gate
18 成贞门　Chengzhen Gate
19 三座门　Triple Gate
20 东天门　East Heavenly Gate
21 内西天门　Inner West Heavenly Gate
22 北天门　North Heavenly Gate
23 外西天门　Outer West Heavenly Gate

1 大祀殿　Hall of Great Sacrifices
2 神厨库　Divine Kitchen and Divine Storeroom
3 宰牲亭　Pavilion of Immolation
4 具服殿　Imperial Dressing Hall
5 斋宫　Palace of Abstinence
6 圜丘　Circular Mound Altar
7 泰神殿　Hall of Imperial Portraits and Tablets
8 神厨库　Divine Kitchen and Divine Storeroom
9 三库　Triple Storeroom
10 宰牲亭　Pavilion of Immolation
11 具服台　Imperial Dressing Platform
12 崇雩坛　Altar of Prayer for Rain
13 神乐观　Divine Music Monastery
14 牺牲所　Building for Sacrificial Livestock
15 祠祭署　Sacrificial Administration
16 昭亨门　Zhaoheng Gate
17 泰元门　Taiyuan Gate
18 广利门　Guangli Gate
19 成贞门　Chengzhen Gate
20 三座门　Triple Gate
21 东天门　East Heavenly Gate
22 内西天门　Inner West Heavenly Gate
23 北天门　North Heavenly Gate
24 外西天门　Outer West Heavenly Gate

明嘉靖十八年（1539年）改建皇穹宇

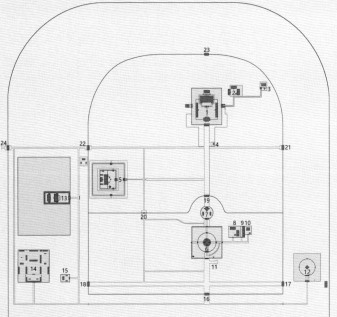

明嘉靖二十四年（1545年）改建大享殿

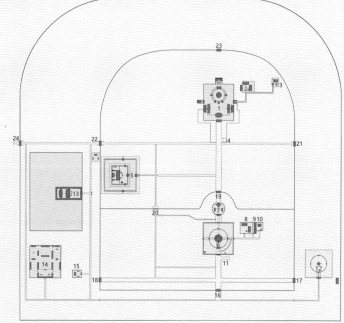

1 大祀殿　Hall of Great Sacrifices
2 神厨库　Divine Kitchen and Divine Storeroom
3 宰牲亭　Pavilion of Immolation
4 具服殿　Imperial Dressing Hall
5 斋宫　Palace of Abstinence
6 圜丘　Circular Mound Altar
7 皇穹宇　Imperial Vault of Heaven
8 神厨库　Divine Kitchen and Divine Storeroom
9 三库　Triple Storeroom
10 宰牲亭　Pavilion of Immolation
11 具服台　Imperial Dressing Platform
12 崇雩坛　Altar of Prayer for Rain
13 神乐观　Divine Music Monastery
14 牺牲所　Building for Sacrificial Livestock
15 祠祭署　Sacrificial Administration
16 昭亨门　Zhaoheng Gate
17 泰元门　Taiyuan Gate
18 广利门　Guangli Gate
19 成贞门　Chengzhen Gate
20 三座门　Triple Gate
21 东天门　East Heavenly Gate
22 内西天门　Inner West Heavenly Gate
23 北天门　North Heavenly Gate
24 外西天门　Outer West Heavenly Gate

1 大享殿　Hall of Great Offerings
2 神厨库　Divine Kitchen and Divine Storeroom
3 宰牲亭　Pavilion of Immolation
4 具服台　Imperial Dressing Platform
5 斋宫　Palace of Abstinence
6 圜丘　Circular Mound Altar
7 皇穹宇　Imperial Vault of Heaven
8 神厨库　Divine Kitchen and Divine Storeroom
9 三库　Triple Storeroom
10 宰牲亭　Pavilion of Immolation
11 具服台　Imperial Dressing Platform
12 崇雩坛　Altar of Prayer for Rain
13 神乐观　Divine Music Monastery
14 牺牲所　Building for Sacrificial Livestock
15 祠祭署　Sacrificial Administration
16 昭亨门　Zhaoheng Gate
17 泰元门　Taiyuan Gate
18 广利门　Guangli Gate
19 成贞门　Chengzhen Gate
20 三座门　Triple Gate
21 东天门　East Heavenly Gate
22 内西天门　Inner West Heavenly Gate
23 北天门　North Heavenly Gate
24 外西天门　Outer West Heavenly Gate

Fig.8 Evolution of the Temple of Heaven in *Shuntian* in the reign of the Jiajing Emperor (Drawn by author)

the two temples were located on the same side of the axis. As for scales, the perimeter of the altar of the Temple of Heaven and Earth was 10 *li* (5760 meters), much longer than the perimeter of the altar in Yingtian, which was 9 *li* and 30 *bu* (5232 meters) (fig.6). Finally, regarding the ambient environments, compared to the Hall of Great Sacrifices in Yingtian, which was located on the site of the Circular Mound Altar (*Yuanqiu*), the Hall of Great Sacrifices inside the Temple of Heaven and Earth in Shuntian was situated on a lofty platform, which impelled people to climb up slowly on the slopes and therefore fed people with senses of sacredness and dignity during their access to gods (fig.7).

2. The Temple of Heaven in the Reign of the Jiajing Emperor

A hundred years after ZHU Di's urban renewal program in Shuntian (modern Beijing), ZHU Houcong, a collateral relative of the imperial family, took the throne of the Ming Dynasty and declared himself the Jiajing Emperor officially in 1522. Due to the fact that ZHU Houcong was not a direct descendant of the former emperor, an exclusively influential and prolonged conflict arose between the newly enthroned emperor and his officialdom, which was known as the Great Rites Controversy (*Daliyi*). The Great Rites Controversy lasted from the year before ZHU Houcong's enthronement to the seventeenth year of his reign (1521-1538). At first, the conflict focused primarily on the identity between the main lineage of the ruling house and the imperial succession. Soon after ZHU Houcong came up with the idea to posthumously elevate his biological father to the status of emperor, the arguments on rituals were put on the table. By the end of the Great Rites Controversy, the majority of the officials were forced to agree that rituals performed contrary to the emperor's own heart would be against human nature. Consequently, extensive modifications were made in both sacrificing rituals and architectural designs of the temples in Shuntian. The Temple of Heaven and Earth (*Tianditan*)—later renamed the Temple of Heaven (*Tiantan*)—was precisely one of these temples with considerably changes. From the third year of the reign of the Jiajing Emperor (1524) when the arguments of rituals were just proposed, to the seventeenth year (1538) when the Great Rites Controversy was ended, the evolution of the Temple of Heaven in Shuntian could be divided into two periods: the period of constructions, in which the major events were the construction of the Circular Mound Altar (*Yuanqiu*) and the construction of the Altar of Prayer for Rain (*Chongyutan*); and the period of reconstructions, in which the reconstruction of the Imperial Vault of Heaven (*Huangqiongyu*) and the reconstruction of the Hall of the Great Offerings (*Daxiangdian*) were the milestones (fig.8).

2. 明嘉靖朝天坛

在北京城与诸坛庙建成百年之后，明嘉靖皇帝朱厚熜以旁系入京赓承大统，为了正宗统、孝皇考，在正德十六年至嘉靖十七年（1521—1538年）间展开了一场规模巨大、旷日持久的礼制争论和政治斗争，史称『大礼议』，与之关系最为密切的北京坛庙也因此迎来了大规模的格局调整。与宗庙同等重要的郊坛，尤其是天坛，以嘉靖十七年（1538年）九月『明堂大享礼』举行即『大礼议』结束为节点，可分为『创建』与『改建』两个时期，『鼎建圜丘』『增设崇雩坛』『改建皇穹宇』『改建大享殿』四个阶段（图八）。

1）鼎建圜丘（1530—1531年）

嘉靖九年（1530年），世宗朱厚熜恢复太祖洪武初制，分祀天地于南北两郊，在天地坛大祀殿之南新建圜丘（图九），复行冬至郊天之礼，『天地坛』更名为『天坛』。翌年，圜丘以北的神御版殿也建成，嘉靖帝定名『泰神殿』。圜丘与泰神殿外有坛墙围合，共设四门，以『乾卦』之『四德』命名，分别为南门『昭亨』、北门『成贞』、东门『泰元』、西门『广利』。

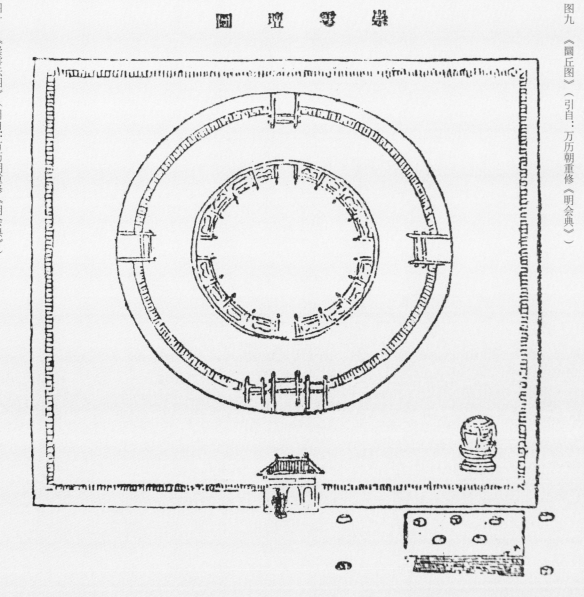

崇雩坛图

图九　《圜丘图》（引自：万历朝重修《明会典》）

圜丘图

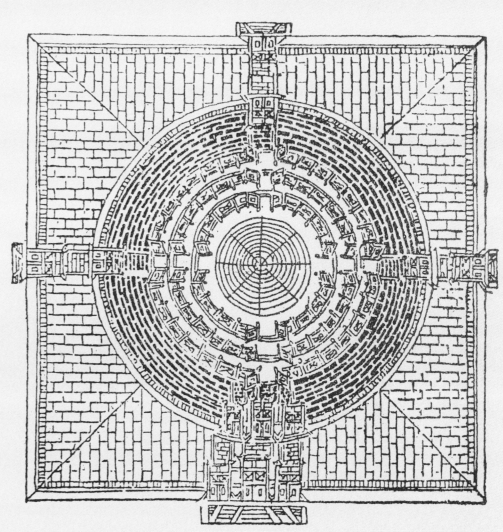

Fig.9 *Map of the Circular Mound Altar* (Source: *Code of the Ming Dynasty*)
Fig.10 *Map of the Altar of Prayer for Rain* (Source: *Code of the Ming Dynasty*)

1) Construction of the Circular Mound Altar (1530-1531)

In the ninth year of his reign (1530), ZHU Houcong, the Jiajing Emperor, changed the rituals for sacrifices from worship heaven and earth together, which was stipulated by ZHU Yuanzhang, back to offer separate sacrifices to heaven and earth. Consequently a new Circular Mound Altar (*Yuanqiu*) was built at the south of the Hall of Great Sacrifices (*Dasidian*) in the Temple of Heaven and Earth in Shuntian to sacrifice heaven at the Winter Solstice (fig.9), and the Temple of Heaven and Earth was renamed the Temple of Heaven since it no longer served earth. The next year (1531), the Hall of Imperial Portraits and Tablets (*Taishendian*) at the north of the Circular Mound Altar was completed. Additional wall was also constructed to enclose the Circular Mound Altar and the Hall of Imperial Portraits and Tablets. The wall had four gates in four directions, which were named after the heaven hexagram—*Qian*—of the *I Ching*, namely Zhaoheng Gate at the south, Chengzhen Gate at the north, Taiyuan Gate at the east and Guangli Gate at the west.

2) Construction of the Altar of Prayer for Rain (1532)

In the early Ming Dynasty, the sacrificial rites for rain during flood and drought were stipulated without specified altars. As a result, when rites were needed, the emperors would either pray in the Forbidden City by themselves, or send officials praying in other spirits' altars, such as the Altar of Land and Grain (*Shejitan*) or the Temple of Mountains and Rivers (*Shanchuantan*). In the first month of the eleventh year of his reign (1532), ZHU Houcong issued the edict to build the Altar of Prayer for Rain (*Chongyutan*) at the north of the Taiyuan Gate of the Circular Mound Altar in the Temple of Heaven, following the rituals of sacrificing in suburbs (fig.10). However, after the completion, the Altar of Prayer for Rain was used only once by ZHU Houcong in the fourth month of the seventeenth year of his reign (1538), and was soon replaced by the Circular Mound Altar as the altar for rain. In the twelfth year of the reign of the Qianlong Emperor in the Qing Dynasty (1747), after being abandoned for almost two hundred years, the Altar of Prayer for Rain was ultimately pulled down.

3) Reconstruction of the Imperial Vault of Heaven (1539-1540)

On the Winter Solstice of the seventeenth year of the reign of the Jiajing Emperor (1538), before offering a generous sacrifice to heaven, ZHU Houcong decided to change the title

2）增设崇雩坛（1532年）

明初定雩祭，为水、旱灾时之祭祀，但并未专设坛壝，或躬祷或露告于宫中，或遣官祭告郊、庙、陵寝及社稷、山川。嘉靖十一年（1532年）正月，诏于圜丘东泰元门外之北，依郊坛之制建崇雩坛（图十）。崇雩坛建成后一直未被使用，直至嘉靖十七年（1538年）四月，嘉靖帝才亲往祭祀。其后明清两代雩祀之礼又皆行于圜丘，至清乾隆十二年（1747年）天坛围垣修整之际，废弃已久的崇雩坛组群被拆除。

3）改建皇穹宇（1539—1540年）

嘉靖十七年（1538年）冬至大祀圜丘前，嘉靖帝更定上帝泰号，尊称为『皇天上帝』，拉开了泰神殿改建皇穹宇的序幕。正式兴工在嘉靖十八年（1539年）八月，至次年七月皇穹宇组群工成（图十一），其主体建筑皇穹宇为周围八柱的重檐圆殿，东西配殿各五间，其外有圆形围墙环绕，南门三间。组群中心北移至原成贞门位置，成贞门亦北移，两侧砌以弧墙与原直墙连接，弧墙与直墙城砖上的窑造款识年代不同正表明经过前后两个时期的营造过程（图十二、图十三）。

Fig.11 *Map of the Imperial Vault of Heaven* (Source: *Code of the Ming Dynasty*)
Fig.12 Inscription Marks on Bricks of Straight Walls of Chengzhen Gate (Photographed by Author)
Fig.13 Inscription Marks on Bricks of Curved Walls of Chengzhen Gate (Photographed by Author)

皇穹宇图

图十一 《皇穹宇图》（引自：万历朝重修《明会典》）

中国古建筑测绘大系·坛庙建筑——天坛

016

图十三 成贞门两侧弧墙城砖窑造款识（作者自摄）

图十二 成贞门两侧直墙城砖窑造款识（作者自摄）

of the God of Heaven from *Shangdi* to *Huangtian Shangdi* and to transform the Hall of Imperial Portraits and Tablets (*Taishendian*) built seven years before into the Imperial Vault of Heaven (*Huangqiongyu*). The transformation began in the eighth month of eighteenth year of the reign of the Jiajing Emperor (1539) and was accomplished in the seventh month of the next year (1540) (fig.11). The new Imperial Vault of Heaven consisted a hall with a round double-eave roof and eight columns at the north, two five-bay side halls at the east and west and a circular wall enclosing the above buildings. There were three gates in the southern part of the circular wall serving as entrances. Due to the changes of the layout, the Chengzhen Gate outside the former Hall of Imperial Portraits and Tablets had to be pushed northwards to make space for the current Imperial Vault of Heaven, while the straight walls connecting to the gate were partly torn down and replaced by curved walls to adapt the change of the Chengzhen Gate. The reconstruction of the walls can be proved by the different inscription marks on the bricks of straight walls and the bricks of curved walls (fig.12, fig.13).

4) Reconstruction of the Hall of Great Offerings (1540-1545)

After the Circular Mound Altar was completed inside the Temple of Heaven, ZHU Houcong stipulated that prayers for harvests should be held on the first day of *Xin* in the first lunar month annually in the Hall of Great Sacrifices in the Temple of Heaven. However, shortly after the first and only prayer held on the day of *Xin* in the first lunar month in the nineteenth year of the reign of the Jiajing Emperor (1531), the prayers for harvests were modified to be held on the Day of Waking of Insects (*Jingzhe*), the fifth or sixth day of the third month, in the Circular Mound Altar. Consequently the Hall of Great Sacrifices was left disused for the following seven years. In the seventeenth year of the reign of the Jiajing Emperor (1538), to follow the ancient rituals of holding harvest prayers in autumn in the Hall of Enlightenment (*Mingtang*), ZHU Houcong, the Jiajing Emperor, issued the imperial edict to hold a prayer for harvests in the Hall of Great Sacrifices. For some reasons, the prayer was eventually held in the Hall of Mysterious (*Xuanjibaodian*) in the Forbidden City. In the tenth month of the nineteenth year of the reign of the Jiajing Emperor (1540), with the same intention as two years before, ZHU Houcong finally decided to demolish the Hall of Great Sacrifices (*Dasidian*), to build the Hall of Great Offerings (*Daxiangdian*) in accordance with the Hall of Enlightenment. The new hall was built on the site of the previous Hall of Great Sacrifices. However, the plan of the round hall was too large that it took over the entire platform and left no space for other buildings. Hence a separate platform was constructed at the north of the

4）改建大享殿（1540—1545 年）

自南郊圜丘建成之后，即议定在每岁春正月上辛日在大祀殿举行祈谷礼，但仅于嘉靖十年（1531 年）正月举行过一次，后即改为启蛰日行祈谷礼于圜丘，大祀殿即废弃不用。至嘉靖十七年（1538 年），世宗欲行『明堂秋享之礼』，位于圜丘之北、皇城东南的大祀殿即成为祀典绝佳之选，但由于诸多原因，大享礼行于大内玄极宝殿。至嘉靖十九年（1540 年）十月，嘉靖帝准古明堂之义，下诏在原大祀殿址建大享殿，而次年的太庙火灾使大享殿工程暂停。停工期间，嘉靖帝下诏拆撤了旧大祀殿。大享殿再次兴工已是嘉靖二十二年（1543 年）三月，其余工程兴工于十月，包括了新建大享神御殿（即皇乾殿）、改建大祀殿东西庑、修缮大祀殿门与东西南三座砖门、拆除具服殿等。至嘉靖二十四年（1545 年）八月，这座准古肇建的鸿构组群即将完工，大臣们奏请当年在大享殿举行秋享上帝之礼，然而始料未及的是，嘉靖帝暂止了与之相关的各项礼仪，仍将行礼于玄极宝殿，且终嘉靖一朝，明堂秋享之礼亦未在大享殿举行。此次新建大享殿即在旧大祀殿原址，由于圆形平面已叠压原天库，因此须在原方形大台基北部加筑，突出高台，容纳皇乾殿，亦形成与皇穹宇组群类似的独立院落（图十四）。

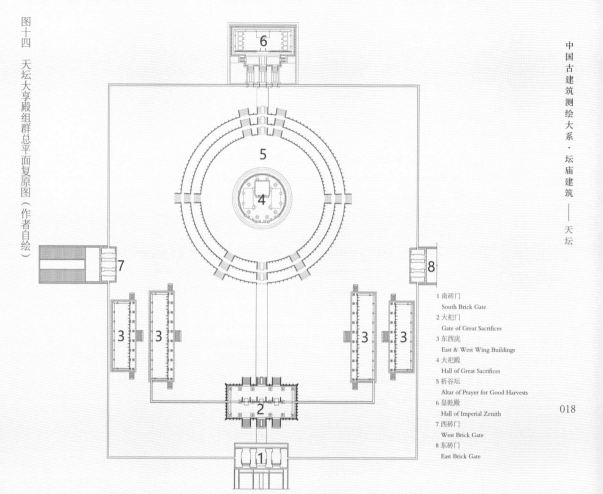

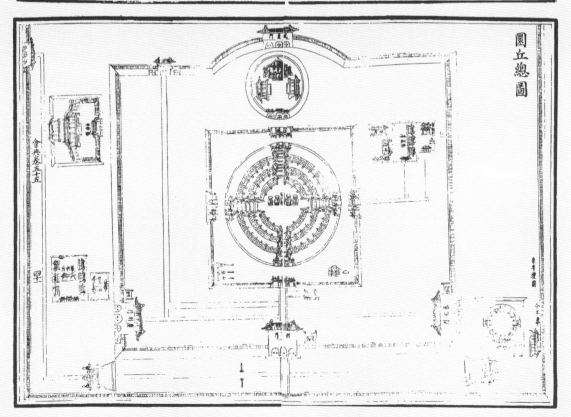

图十五　祈谷坛图（上图）与圜丘总图（下图）（引自：康熙朝《大清会典》）

图十四　天坛大享殿组群总平面复原图（作者自绘）

1 南砖门
  South Brick Gate
2 大祀门
  Gate of Great Sacrifices
3 东西庑
  East & West Wing Buildings
4 大祀殿
  Hall of Great Sacrifices
5 祈谷坛
  Altar of Prayer for Good Harvests
6 皇乾殿
  Hall of Imperial Zenith
7 西砖门
  West Brick Gate
8 东砖门
  East Brick Gate

Fig.14 Restored Map of the Complex of Hall of Great Offerings
Fig.15 Map of the Hall of Prayer for Good Harvests (the first figure) and Map of the
Circular Mound Altar (the second figure) (Source: *Code of the Great Qing Dynasty*)

former platform to accommodate the Hall of Imperial Zenith (*Huangqiandian*), making it an isolated complex as the Imperial Vault of Heaven (fig.14). In the next year (1541), the construction of the new hall was temporarily stopped by a conflagration started in the Imperial Ancestral Temple (*Taimiao*), but the demolition of the old buildings was kept carrying. Two years later (1543), the construction of the Hall of Great Offerings was resumed, as well as the construction of the Hall of Imperial Zenith, the renovation of the wing buildings, the restoration of the Gate of Great Sacrifices (*Dasimen*) and three brick gates at the south, west and east, and the demolition of the Imperial Dressing Hall (*Jufudian*). In the eighth month of the twenty-fourth year of the reign of the Jiajing Emperor (1545), the grand complex was almost completed. Thus, the officials proposed a prayer for harvests in the upcoming autumn of that year in the Hall of Great Offerings as celebrations. ZHU Houcong, however, unexpectedly declined and held the cult in the Hall of Mysterious instead. Until the end of the reign of ZHU Houcong, no more prayers were held in the Hall of Great Offerings.

III. The Qing Dynasty: The Evolution of the Temple of Heaven

In the evolution history of the Temple of Heaven (*Tiantan*) in the Qing Dynasty, there are three milestones that should be mentioned, that is, the establishment of the ritual system during the reign of the Shunzhi Emperor, the grand reformation in the reign of the Qianlong Emperor and the reconstruction of the Hall of Prayer for Good Harvests (*Qiniandian*) in the reign of the Guangxu Emperor.

1. Establishment of the Ritual System in the Shunzhi Emperor's Reign

After the conquest of the Ming, the Qing Dynasty inherited *Shuntian* as the capital and therefore adopted the imperial palaces, temples and altars in *Shuntian* from the Ming Dynasty, the Temple of Heaven (*Tiantan*) included. The next two emperors after Emperor Shunzhi hardly did construction or reconstruction to it. [①] At that time, the Temple of Heaven was in a pretty stable condition due to the constructions during the reign of the Jiajing Emperor. The Circular Mound Altar (*Yuanqiu*), the Hall of Great Offerings (*Daxiangdian*), the Altar of Prayer for Rain (*Chongyutan*), the Palace of Abstinence (*Zhaigong*), the Divine Music Monastery (*Shenyueguan*) and the Building for Sacrificial Livestock (*Xishengsuo*) were all well preserved from the changing of dynasties (fig.15).

三、清北京天坛

纵观北京天坛在清代的历史沿革与建筑变迁，其中有三个时段尤为重要，即顺治朝的祀典创立、乾隆朝的兴修改易与光绪朝的祈年殿重建。

1 顺治朝的祀典创立

清顺治帝福临定鼎燕京，宫室、宗庙、郊坛多沿用前明旧物，天坛建筑更因明嘉靖朝奠定的宏格局，直至康熙、雍正两朝都未有重要的营缮活动。[①]而顺治朝正值国祚未稳、百废待兴之际，面对明代天坛遗留下来的圜丘、大享殿、崇雩坛、斋宫、神乐观、牺牲所六大组群（图十五）以及纷繁复杂的祀典礼仪更旋变革，则大清王朝的祀天礼仪创立尤为重要。

顺治元年（1644年）十月初一，顺治帝以定鼎燕京亲诣南郊，告祭天地即皇帝位，次月即定制每岁冬至大祀天于圜丘。十三年（1656年）又议定恢复祈谷之礼，以每岁正月上辛日，恭祀上帝于祈谷坛即大享殿下三层圆台。十七年（1660年）议定每年孟春合祭天地日月及诸神于大享殿，

In fact, there was hardly any restoration within the temple in the first hundred years of the Qing Dynasty. Though the architecture remained intact, the rituals for sacrificing were still indefinite due to the chaos caused by wars. Under this condition, as the first emperor of the Qing Dynasty, Fulin was eager to reform the ritual system to show his domination of the country.

On the first day of the tenth month of the first year of his reign (1644), Fulin, the Shunzhi Emperor, offered a sacrifice in the Temple of Heaven in order to gain his legitimacy from heaven and earth. The next month, he announced that sacrifices to heaven should be held at the Winter Solstice annually in the Circular Mound Altar. In the thirteenth year of his reign (1656), he decided to resume the cult of holding prayers for grains annually on the first day of *Xin* of the first month on the three-layer circular platform of the Hall of Great Offerings. In the seventeenth year of his reign (1660), Fulin declared that each year, a comprehensive sacrifice should be offered in the Hall of Great Offerings in the first month of spring, in which heaven, earth, sun, moon and all spirits would be worshipped. Nevertheless, until it was abolished in the reign of the next emperor, only one sacrifice was offered in the fourth month of the same year he gave the instruction. In addition to the above rituals, there was an unwritten rule started by Fulin in the fourteenth year of his reign (1657) that the emperor would pray for rain in the Circular Mound Altar during drought. It was formalized in the seventh year of the reign of the Qianlong Emperor (1742).

2. Reformation of the Temple of Heaven in the Qianlong Emperor's Reign

The development of the Qing Dynasty reached to its peak in the reign of Hongli, the Qianlong reign. At that time, the buildings inside the Temple of Heaven were in their worst conditions after a hundred years' deteriorations: the walls were collapsed, the colored paintings were faded, the layouts failed to match the emperor's ambitions, the rites were in chaos. Therefore, from the seventh to the twentieth year of Hongli's reign (1742-1755), a grand reformation was carried out on the buildings and structures in the Temple of Heaven (*Tiantan*).

From the seventh to the eighth year of the reign of Hongli, the Qianlong Emperor (1742-1743), the Palaces of Abstinence (*Zhaigong*) in the southern and northern suburbs were both restored. The restoration of the palace in southern suburb, the one inside the Temple of Heaven, consisted of two periods. The first period was to restore the main halls of

但仅于当年四月举行了一次合祀大享礼，康熙朝遂废止。此外，自顺治十四年（1656年）始，如遇大旱，皇帝则前往圜丘祷雨，至乾隆七年（1742年）圜丘雩祭之礼成为定制。

2. 乾隆朝的兴修改易

乾隆皇帝在位期间，清朝达到鼎盛，而经阅久远的天坛建筑，或年久倾圮，或金碧不鲜，或空间局促，或取义未协。自乾隆七年至乾隆二十年（1742—1755年），天坛诸多组群以及内外坛垣皆进行了大规模的兴修改易。

乾隆七年至八年（1742—1743年）修理南北两郊斋宫，此次天坛斋宫工程主要包括两部分：其一，修理了明代的正殿、钟楼、铜人亭、宫门、围廊、桥梁、河道；其二，拆除了明代的后殿与配殿，填埋内护城河西段以使内宫墙向西扩展，形成独立的寝宫院落，新建内宫门、正殿、配殿及连廊。嘉庆十二年（1807年）十一月，因熏炕致使寝宫正殿与配殿失火，嗣后在原址重建正殿，但取消了配殿与连廊，正殿后部左右添建随事房各五间，并增加垣墙，对寝宫院进行合理分隔，形成现今的格局。

the Palace of Abstinence, the Bell Tower (*Zhonglou*), the Pavilion of the Bronze Statue (*Tongrenting*), as well as gates, corridors, bridges and channels of the palace left by the Ming Dynasty. The second period was to build the Complex for Recess (*Qingong*) in the Palace of Abstinence, in which the emperors stayed overnight during abstinence. To do this, the rear and side halls of the Palace of Abstinence were first pulled down. Then the western part of the palace moat was filled in so that the frontier of the palace could be pushed westwards, leaving an empty space inside the palace. Finally, the Complex for Recess was built in that space. The Complex for Recess consisted of a main hall, several side halls, corridors and gates. The main and side halls of the complex were burnt down in the eleventh month of the twelfth year of the reign of Hongli's son (1807) in a fire set by the firewood in the furnace. After the fire, the main hall was rebuilt at the original site while the side halls and corridors were knocked down. Two five-bay maid rooms were added at the northeast and northwest of the main hall and several walls were built for comfort. Then the layout of the current Palace of Abstinence was formed.

In the fourth month of the eighth year of the Qianlong Emperor's reign (1743), the Divine Music Monastery (*Shenyueguan*) in the Temple of Heaven was renamed the Divine Music Institute (*Shenyuesuo*) to eliminate the influence of Taoism. The tablets of Taoist spirits inside were also removed except for the tablet of the Taoist Lord of Heaven (*Xuantianshangdi*).

In the twelfth year of the reign of the Qianlong Emperor (1747), both the inner wall and the outer wall of the Temple of Heaven were restored. In the past, the walls of the Temple of Heaven were made up of soil. While during the restoration, bricks were attached outside the soil walls. What's more, the corridors on both sides of the inner wall and the Altar of Prayer for Rain (*Chongyutan*), which was constructed initially in the reign of the Jiajing Emperor of the Ming, were all pulled down.

In the fifth month of the fourteenth year of his reign (1749), Hongli, the Qianlong Emperor, had a discussion with his grand secretary on the restorations of the altars in the southern and northern suburbs during his abstinence. The discussion eventually led to the grand reformation of the core complexes of the Temple of Heaven, which were instructed by one noble and three officials: namely the Prince He (Hongzhou), the brother of the emperor; the Minister of Finance (Haiwang); the Minister of Construction (Sanhe); and the previous Minister of Rites (WANG Anguo). The reformation began with the Hall of Great Offerings (*Daxiangdian*) in the fifteenth year of the Qianlong Emperor's reign (1750). Two years later (1752), the renovation of the Circular Mound Altar (*Yuanqiu*)

八年（1743年）四月，天坛神乐观更名为『神乐所』，仅保留原显佑殿供奉的玄天上帝，将其余道教神灵皆撤去。

十二年（1747年）修理天坛内外坛墙，对原有土墙外包砌城砖，并取消内坛墙两侧出廊。同时，拆除了圜丘泰元门外明嘉靖朝营建的崇雩坛组群。

十四年（1749年）五月，乾隆帝在北郊祭地斋戒时与大学士合议南、北郊坛修缮事宜，由此拉开了天坛核心建筑组群的改易序幕，任命和亲王弘昼、户部尚书海望、工部尚书三和与曾任礼部尚书的王安国为总理工程大臣，十五年（1750年）兴工。首先开工的是祈谷坛大享殿组群，两年后圜丘与皇穹宇组群开工，至十八年（1753年），天坛工程工竣。此次大修对核心建筑的改易主要有以下几个方面：

1）明代皇乾殿为青色瓦顶，其周围围垣及门的瓦顶颜色为绿色，此次修缮后，皇乾殿院落瓦顶颜色统一成青色。

2）原大享殿更名为『祈年殿』，瓦顶颜色由绿色改为青色；东西两庑则拆除了后重的七间，仅保留前部的九间，并改绿色瓦顶为青色。南、东、西三座砖门及坛外垣瓦顶颜色仍为绿色。

大享门更名为『祈年门』，瓦顶颜色由原来的上青、中黄、下绿改为三层皆覆青瓦；原

3）明代圜丘三层面径分别为五丈九尺、九丈、十二丈，通体青色琉璃。此次乾隆帝以『九五之数』为原则扩建圜丘，设计方案经多轮修改，最终按康熙朝《律吕正义》所稽考的古尺○为度，三层直

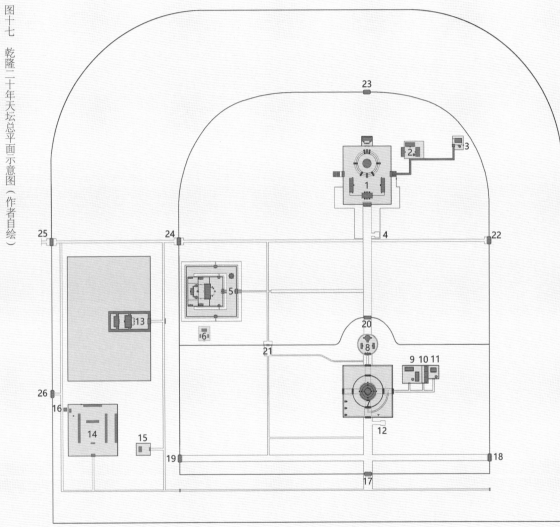

图十七　乾隆二十年天坛总平面示意图（作者自绘）

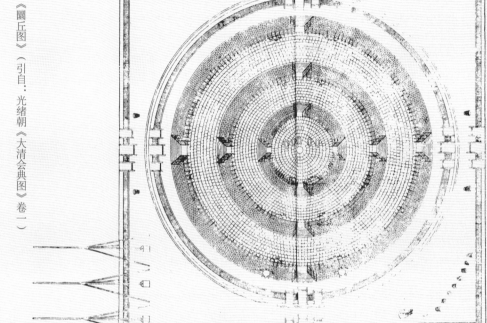

图十六　《圜丘图》（引自：光绪朝《大清会典图》卷一）

1 祈年殿
Hall of Prayer for Good Harvests
2 神厨库
Divine Kitchen and Divine Storeroom
3 宰牲亭
Pavilion of Immolation
4 具服台
Imperial Dressing Platform
5 斋宫
Palace of Abstinence
6 祠祭署
Sacrificial Administration

7 圜丘
Circular Mound Altar
8 皇穹宇
Imperial Vault of Heaven
9 神厨库
Divine Kitchen and Divine Storeroom
10 三库
Triple Storeroom
11 宰牲亭
Pavilion of Immolation
12 具服台
Imperial Dressing Platform

13 神乐观
Divine Music Administration
14 牺牲所
Building for Sacrificial Livestock
15 祭器库
Sacrificial Utensil Storeroom
16 钟楼
the Bell Tower
17 昭亨门
Zhaoheng Gate
18 泰元门
Taiyuan Gate

19 广利门
Guangli Gate
20 成贞门
Chengzhen Gate
21 三座门
Triple Gate
22 东天门
East Heavenly Gate
23 北天门
North Heavenly Gate
24 内西天门
Inner West Heavenly Gate

25 外西天门
Outer West Heavenly Gate
26 圜丘门
Gate of Circular Mound Altar

Fig.16 *Map of the Circular of Mound Altar* (Source: *Code of the Great Qing Dynasty* vol.1)
Fig.17 Map of the Temple of Heaven in the twentieth year of Qianlong Emperor's Reign (1755)

and the Imperial Vault of Heaven (*Huangqiongyu*) were started as well. In the eighteenth year of the reign of the Qianlong Emperor (1753), the grand reformation of the Temple of Heaven was completed. The vital changes in the reformation were listed as follows:

1) Renovation of the Hall of Imperial Zenith (*Huangqiandian*). In the Ming Dynasty, the roof of the main hall of the Hall of Imperial Zenith was paved with ultramarine glazed tiles, while the roofs of the surrounding walls and gates were paved with green glazed tiles. After the renovation in the reign of the Qianlong Emperor, all roofs belonging to the Hall of Imperial Zenith were re-paved with ultramarine glazed tiles.

2) Reformation of the Hall of Great Offerings. In the Ming Dynasty, the three eaves of the Hall of Great Offerings were paved with ultramarine, yellow and green glazed tiles from the top to the bottom, while the eave of the Gate of Great Offerings (*Daxiangmen*) was paved with green glazed tiles. After the grand reformation, all three eaves of the hall and the eave of the gate were re-paved with ultramarine tiles. Meanwhile, the hall and the gate were renamed the Hall of Prayer for Good Harvests (*Qiniandian*) and the Gate of Prayer for Good Harvests (*Qinianmen*), the same names as they are still using now. As for the four wings at the eastern and western sides of the Hall of Prayer for Good Harvests, two of them stretching to the north were demolished and two stretching to the south were kept. The roofing tiles of the remaining wings were changed from green into ultramarine. In contrast, the green glazed roofing tiles of the southern, eastern, western brick gates and the outer wall were remained.

3) Enlargement of the Circular Mound Altar. In the Ming Dynasty, the diameters of the three tiers of the altar inside the Circular Mound Altar were 59 *chi* (18.9 meters), 90 *chi* (28.8 meters) and 120 *chi* (38.4 meters) respectively from the top to the bottom. Unsatisfied with the scale of the altar, Hongli decided to enlarge the Circular Mound Altar based on the Principle of Nine and Five in the grand reformation. After several revisions to the plan, the final decision was made to enlarge the top tier of the altar to 90 *chi* (28.8 meters) in diameter, the middle tier to 150 *chi* (38.4 meters) in diameter and the bottom tier to 210 *chi* (67.2 meters) in diameter. The enlargement was made according to the dimension standards in *Imperial Standard Interpretation of Tone-System (Yuzhi Lvlv Zhengyi)* [2] . Besides the scale, the materials used for the altar were also changed, from ultramarine glaze slabs to grey limestone slabs. On each tier of the altar there were nine circles of limestone slates on the floor. The edges of the altar were engraved with lotus patterns. Three hundred and sixty—the number of the circumference of heaven—

径分别为九丈、十五丈、二十一丈；坛面材质为艾叶青石，三层各九重石块，仰覆莲须弥座，上安360楹青色琉璃栏板（图十六）。建造过程中，因琉璃栏板难以烧造，改为三层艾叶青石栏板，数量也减少至『乾策之数』的216楹，并添安龙头。此外，圜丘坛外的圆方两重墙墙瓦顶颜色也由绿色改为青色。

4) 此次皇穹宇改建工程，把嘉靖朝的重檐圆殿改为了单檐圆殿，以符『阳数』；原有青色琉璃槛墙加上身青灰墙做法改为临清城砖干摆做法；台面原有一圈青色琉璃砖墁地也改为青白石墁地。此外，围垣及门的瓦顶颜色也由绿色改为青色。

至此，乾隆朝复鼎新南郊坛宇，凡崇卑之制，象色之宜，无不斟酌尽善。

此后，乾隆十九年（1754年）先是修理了牺牲所，对原有建筑布局及神所、官廨、牲房等的位置与间数进行了调整；又于天坛西外垣之南，与先农门相对处增建一座坛门，并于外垣内增建钟楼一座，此后遇祭圜丘即由此新建的圜丘门进入，祭祈谷坛则仍由原有的外西天门进入。二十年（1755年），神乐所又改名为『神乐署』（图十七）。

3. 光绪朝的祈年殿重建

光绪十五年（1889年）八月二十四日，祈年殿毁于雷火。祈年殿关乎国家重要祀典，遂议定重建，因无相关图纸，乃集工匠及相关人员询问并进行勘察与设计。十六年（1890年）祈年殿补建工程兴工，至二十五年（1899年）三月工程验收。此次重建的祈年殿与烧毁前的祈年殿相比较，其平面尺度相同，

ultramarine glazed balusters were planned to be placed on the altar (fig.16). However, owning to the difficulty of producing glazed slabs in such complicated shapes, it turned out that two hundred and sixteen—the number of the divination sign of heaven—limestone balusters, with carved dragons' heads on the top, were set instead at last. The roofs of the two layers of walls, the circular inner wall and the square outer wall, outside the altar were re-paved with ultramarine glazed tiles.

4) The Reconstruction of the Imperial Vault of Heaven. In the reconstruction of the Imperial Vault of Heaven, Hongli, the Qianlong Emperor, transformed the double-eave round roof of the main hall into a single-eave round roof, because the number of the eave should represent Yang rather than Yin, altered the slates around the main hall from ultramarine glazed slates into limestone slates, and changed the materials used on walls of the main hall. In the past, the walls of the main hall were a mixture of two materials: the lower parts were made out of ultramarine glazed bricks; the upper parts were made out of gray bricks. After the reconstruction, both parts of the walls were rebuilt with bricks from Linqing, the most famous and high-quality bricks of the time, using the technique called *Ganbai*, the highest bricklaying technique in China. Apart from the main hall, the colors of the glazed roofing tiles of the walls and gates were also altered from green into ultramarine.

After the above changes were made, Hongli, the Qianlong Emperor, resumed sacrifices in the southern suburb. The renewed sacrifices strictly followed the hierarchical system and the ancient tenets relating numbers and colors. The details of the sacrifices were all under the most careful considerations.

In the following years, several minor changes were committed in the Temple of Heaven. In the nineteenth year of the Qianlong Emperor's reign (1754), the Building for Sacrificial Livestock (*Xishengsuo*) was restored: the general layout of the complex was changed, as well as the settings and measurements of the Hall for the God of Sacrificial Livestock (*Shensuo*), the Office for Officials (*Guanxie*) and rooms for various livestock. A new temple gate was built at the southern end of the western outer wall, opposite to the gate of the Temple for the Divine Cultivator (*Xiannongtan*), and was named the Gate of the Circular Mound Altar (*Yuanqiumen*). Compared to the old West Gate (*Xitianmen*) that located at the northern end of the western wall, which was used for rites in the Hall of Prayer for Good Harvests, the new gate was mainly used for sacrifices in the Circular Mound Altar. Moreover, a new Bell Tower was constructed inside the outer wall. In

但在建筑高度、出檐尺度、屋面举折、陡匾尺寸等方面略有区别。值得重视的是，这次工程留有样式雷《天坛补修工程做法册》（图十八），是重修时工部算房的底本，虽在施工中略有更变，但大体仍与现状符合，是研究此殿结构做法的重要档案。

四、现代保护和研究

1900 年八国联军侵华战争爆发，侵华军队侵入北京，英军侵占天坛。《辛丑条约》签订后，天坛又短暂地回归清朝政府管理使用。1911 年辛亥革命爆发，结束了清王朝统治，次年清帝退位，中华民国成立，天坛由清朝典礼院移交中华民国政府内务部管理。1913 年元旦天坛首次对民众短暂开放，5 月《坛庙管理大纲》颁布，北京坛庙分为上中下 3 个等级，天坛为上等级保护文物。

1918 年元旦，天坛正式辟为公园。至此，这座历史悠久的皇家祭坛被赋予了新的功能与内涵。

20 世纪 20 至 30 年代，中国文物古迹的调查研究与保护整理工作日益得到各级政府和学术界的关注。1928 年和 1929 年相继成立的中央古物保管委员会和中国营造学社，开展了一系列的全国性文物保护与研究工作。1935 年 1 月，鉴于北京古物古迹规模大、数量多且多有残损毁圮，旧都文物整理委员会及其执行机构北平文物整理实施事务处宣告成立，负责整理修缮北平文物古迹事

the next year (1755), the Divine Music Institute was renamed as the Divine Music Administration (*Shenyueshu*) (fig.17).

3. Reconstruction of the Hall of Prayer for Good Harvests in the Guangxu Emperor's Reign

On the twenty-fourth day of the eighth month in the fifteenth year of the Guangxu Emperor's reign (1889), a fire caused by lightning hit the Hall of Prayer for Good Harvests (*Qiniandian*) and left the hall in ruins. Since the Hall of Prayer for Good Harvests was closely related to the legitimacy of the country, decision was soon made to rebuild the hall. Due to lack of architectural drawings, the reconstruction did not start until the craftsmen and people involved were called to help in surveying and designing. In the sixteenth year of the reign of the Guangxu Emperor (1890), the reconstruction of the Hall of Prayer for Good Harvests was started. Nine years later (1899), the project was finally completed. Despite the fact that the newly built hall shared the same layout as the burnt hall, the two buildings were different in many aspects, such as the height, the scales and slopes of the three eaves, and the dimensions of the vertical inscribed board on which the name of the hall was. It was worth noticing that a booklet named *Book of Construction Methods in Reconstruction of the Temple of Heaven (Tiantan Buxiu Gongchengzuofa Ce)*, written by Family Lei (*Yangshi Lei*), the most famous family of architects in the Qing Dynasty, was left (fig.18). The booklet was a manuscript used by the accounting office to balance the books. Although a little different from the final reconstruction, the booklet was still an important reference in terms of the structure of the Hall of Prayer for Good Harvests.

IV. Modern times: Research and Preservation

In 1900, the Siege of the International Legations erupted and the Eight-Nation Alliance soon invaded Beijing. The British military occupied the Temple of Heaven (*Tiantan*) and turned it into the force's temporary command in Beijing. In September 1901, the Qing Dynasty and the Eight-Nation Alliance signed the Boxer Protocol (*Xinchou Tiaoyue*) and the Temple of Heaven was returned to the empire for a short time. In 1911, the Xinhai Revolution, a Chinese bourgeois democratic revolution, overthrew the Qing Dynasty. In 1912, the Republic of China was established and took over the Temple of Heaven. The next year, the government opened the Temple of Heaven to the public for the first time from January 1st to 10th. In May, the *Outline of Temple Management (Tanmiao Guanli*

宜。在此背景下，天坛修缮工程成为旧都文物整理一期工程中超前筹划、开工最早、经费列支最多的一项，1935 年 5 月开工，翌年 10 月竣工。在此工程中，由中国营造学社、基泰工程司、恒茂木厂联手形成的学术研究、勘察设计、修缮施工精英团队，使此次天坛修缮工程成为我国文物建筑修缮保护史上浓墨重彩的一笔（图十九）。值得一提的是，1935 年 3 月，单士元即在《中国营造学社汇刊》第五卷第三期发表了《明代营造史料·天坛》一文，搜集整理了有关天坛营造的文献史料，考证了明清两代天坛的历史沿革，成为中国学术界研究天坛之最早成果，同时天坛修缮工程也直接受益于此文。

20 世纪 40 年代，为预防北平古建筑遭战火焚毁，由营造学社社长朱启钤谋划，由基泰工程司张镈主持，带领天津工商学院毕业生，历时四年，分三期，测绘了北起钟鼓楼、南至永定门的北京城中轴线主要古建筑，共绘制实测图 704 幅。其中，天坛测绘应在第三期，从 1944 年 3 月开始，

补建

祈年殿三重檐圓亭一座週圍圓十二間各面濶二丈對徑七丈七尺內週
圍廊各深七尺五寸中檐鑚金廊各深一丈一尺下檐柱高二
丈五寸徑二尺二寸金柱高三丈六尺二寸徑三尺鑚金柱高五丈
七尺四寸徑三尺七寸三重檐十五檩週圍攢尖下檐週圍擺
三寸單翹單昂溜金斗科中檐擺安斗口四寸單翹重昂斗口上
檐擺安斗口四寸重翹重昂斗科內裡中檐博脊枋採做斗口三
寸一斗三升荷葉斗科上檐龍井週圍棨鋪順望板上金脊卖排棨
科承椽枋裡面採做斗口三寸一斗三升雲拱斗科內裡榕井天
花廊內崔督頭停出檐飛方圓棨鋪順望板上金脊卖排棨

图十八 《天坛补修工程做法册》首页（国家图书馆藏）

圜 丘 坛 彩 色 图

总序号：016 —— 天坛 —— 圜丘坛 圜丘坛彩色图

图二十 圜丘立面渲染图（引自：《北京城中轴线古建筑实测图集》）

图十九 圜丘工程开工合影，左一杨廷宝，左三林是镇，右二梁思成，右三刘南策（中国文化遗产研究院藏）

Fig.18 First Page of *Book of Construction Methods in Reconstruction of the Temple of Heaven* (Collected by National Library of China)

Fig.19 Group Photo on the Commencement Day of the Restoration Project of the Circular Mound Altar: *Yang Tingbao* (first on the left), *Lin Shizhen* (third on the left), *Liang Sicheng* (second on the right) and *Liu Nance* (third on the right) (Collected by Chinese Academy of Cultural Heritage)

Fig.20 Elevation of the Circular Mound Altar (Source: *Surveying and Mapping Atlas of Traditional Chinese Architecture on Beijing Central Axis*)

Dagang) was published, in which the temples in Beijing were classified into three levels according to its value. The Temple of Heaven was listed as the highest level. On January 1st, 1918, the Temple of Heaven was turned into a park. This endowed the previous imperial sacrificial monument a new function and a new connotation.

In the 1920s and 1930s, the Chinese government and scholars showed solicitude for the research and preservation of Chinese heritage sites. Under this circumstance, the Central Commission for the Preservation of Antiquities and the Society for the Study of Chinese Architecture were established respectively in 1928 and 1929. The two institutes developed a series of researches and preservation projects all over China. In January 1935, considering the large quantity, the enormous scales, and the damaged conditions of the heritage sites in Beijing, the Beijing Commission for the Preservation of Cultural Relics was set up. This commission was responsible for the preservation and restoration of cultural relics in Beijing, of which the Temple of Heaven was the first batch to be restored—started in May 1935 and completed in October 1936. Being the earliest to plan, the earliest to launch and the most money-consuming project, the restoration of the Temple of Heaven attracted several organizations to study and work together, including the Society for the Study of Chinese Architecture, the Jitai Project Department and the Hengmao Timber Mill. Owning to their efforts, the restoration of the Temple of Heaven has become a milestone in the history of heritage preservation (fig.19). It is worth mentioning that in March 1935, SHAN Shiyuan, a member of the Society for the Study of Chinese Architecture, published the article *the Construction Records of the Temple of Heaven in the Ming Dynasty (Mingdai Yanjiu shiliao, Tian Tan)* in the *Bulletin of Society for Research in Chinese Architecture (Zhongguo Yingzaoxueshe Huikan)*, volume 5, issue 3. The article collected the historical records on the construction of the Temple of Heaven in the Ming and Qing dynasties. It was the first research focusing on the Temple of Heaven and had benefited the restoration project a lot.

In the 1940s, considering that the historical sites in Beijing might be damaged by war, ZHU Qiqian, the president of Society for the Study of Chinese Architecture, proposed an overall surveying and mapping project to record the main historical buildings located on the central axis of Beijing, from the Bell Tower (Zhong Lou) at the northern end to the Yongding Gate at the southern end. The project was led by ZHANG Bo, an architect from Jitai Project Department, and lasted for almost four years. Three batches of graduates from *Institut des Hautes Etudes et Commerciales* in Tianjin took part in the project. The final results included seven hundreds and four drawings. According to records, the

绘制天坛主要建筑的测绘图纸 66 张，其中彩图 6 张（图二十）、墨线图 60 张（图二十一）。与国人测绘天坛同时期，日本学者石桥丑雄在日伪北平特别市公署任职期间，也对天坛进行了调查、测绘与研究，拍摄了大量照片，并绘制了祈年殿、圜丘、皇穹宇、斋宫、南北神厨库几个组群的总平面，回日本后编著《天坛》一书，于 1957 年出版。

中华人民共和国成立后，在政府与全社会的努力下，天坛公园的基础设施得到大幅提升，园林绿化逐步恢复并蓬勃发展，古建保护修缮科学有序进行。20 世纪 80 年代以来，天坛公园管理处作为天坛遗产的守护者，一方面，发挥自身优势开展学术研究，陆续出版了多部天坛研究著作（图二十二）；另一方面，积极与高等院校合作，对天坛古建筑进行系统的测绘。

在此背景下，天坛公园成为天津大学最重要的教学与科研基地。1986 年，天津大学建筑系1984 级学生首次对天坛古建筑进行了探索性的小规模测绘。此后，以天坛申请并获准成为世界文化遗产为契机，1998 年、1999 年，对圜丘、斋宫、神乐署、坛门、北神厨库、皇乾殿等建筑进行

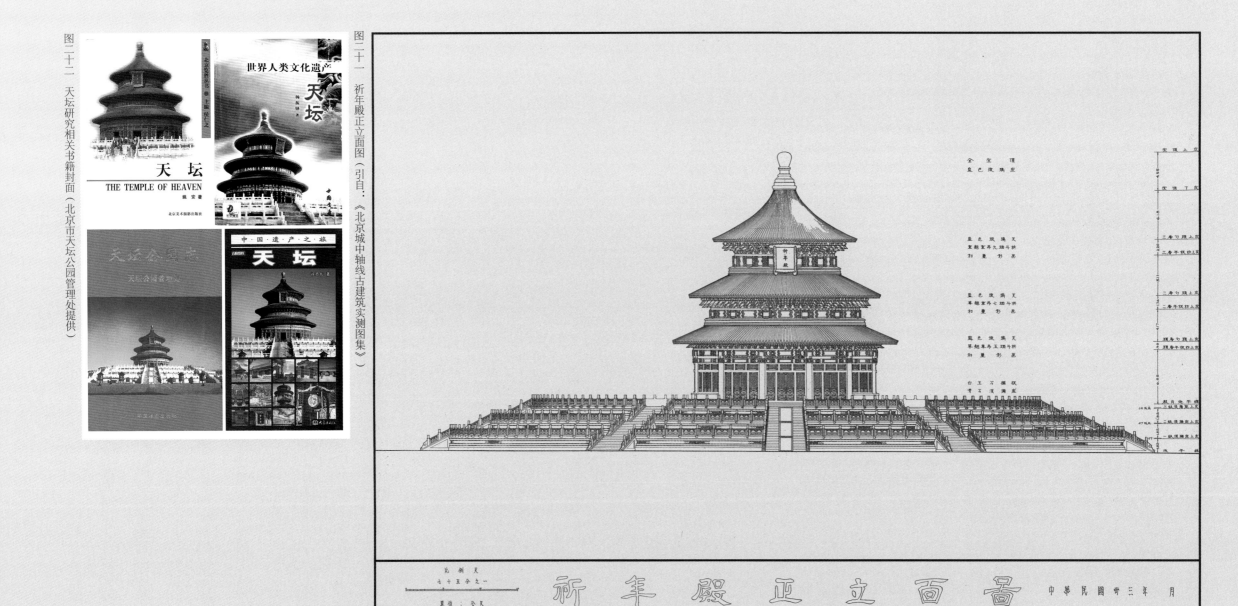

Fig.21 Front Elevation of the Hall of Prayer for Good Harvests (Source: *Surveying and Mapping Atlas of Traditional Chinese Architecture on Beijing Central Axis*)

Fig.22 Covers of Publications on the Temple of Heaven (Source: Photograph Courtesy of the Administration of the Temple of Heaven)

总序号: 051 —— 天坛 —— 祈年殿 —— 祈年殿正立面图

surveying and mapping of the Temple of Heaven was begun in March 1944. It was done by the third batch of graduates from *Institut des Hautes Etudes et Commerciales*. Sixty-six drawings were presented at the end (fig.20) (fig.21). While the graduates were surveying and mapping, Ishibashi Ushio, a Japanese scholar worked for the Puppet Beijing Special Municipal Government Office, was also studying the Temple of Heaven. After minute observation, surveying, mapping and photographing, he managed to draw the master plans of the Hall of Prayer for Good Harvests (*Qiniandian*), the Circular Mound Altar (*Yuanqiu*), the Imperial Vault of Heaven (*Huangqiongyu*), the Palace of Abstinence (*Zhaigong*), the southern and northern Divine Kitchen (*Shenchu*) and the Divine Storeroom (*Shenku*). After returning to Japan, he authored the book *The Temple of Heaven (Tiantan)* in 1957, in which all his researches were published.

In 1949, the People's Republic of China was founded. With the supports of government and society force, the Temple of Heaven—a heritage site and a public green park—was significantly improved in terms of basic facilities, landscaping management and scientific preservation. In the 1980s, the Administration of the Temple of Heaven Park was established and assigned the guardians of the temple. Since then, the administration has devoted into academic researches and has published multiple books about the Temple of Heaven (fig.22). It has also cooperated with universities on surveying and mapping.

Tianjin University is one of most important collaborators of the Administration of the Temple of Heaven Park in terms of surveying and mapping. The first trial collaboration was in 1986, when sophomores from the School of Architecture, Tianjin University tried surveying a small amount of buildings inside the Temple of Heaven. It was after the Temple of Heaven was inscribed the World Heritage and gained much attention that Tianjin University launched the thorough surveying and mapping project. In 1998 and 1999, Tianjin University successively finished the digital surveying and mapping of the Circular Mound Altar (*Yuanqiu*), the Palace of Abstinence (*Zhaigong*), the Divine Music Administration (*Shenyueshu*), the Northern Divine Kitchen (*Shenchu*) and Divine Storeroom (*Shenku*), the Hall of Imperial Zenith (*Huangqiandian*), and altar gates. In 2005 and 2006, with the restoration of the Temple of Heaven, additional surveying was done on the Complex of Hall of Prayer for Good Harvests (*Qiniandian*) and the Complex of Imperial Vault of Heaven (*Huangqiongyu*). With the development of surveying instruments, from 2015 to 2017, the surveyors from Tianjin University mapped the main buildings again with the 3D laser scanner. Since the year 1986, about two hundreds faculties and students from Tianjin University have participated in the surveying and

了大规模的数字化测绘，结合 2005 年、2006 年天坛古建筑大修，对祈年殿组群、皇穹宇组群进行了补测工作；2015 至 2017 年又多次利用三维激光扫描等先进技术对天坛重要建筑进行了复测工作。在此期间，天津大学共有近 200 名师生投入天坛测绘工作中，完成测绘图共计 300 余幅。

为了更好地保护与研究天坛，更形象地展现天坛古建筑，天坛公园管理处和天津大学建筑学院密切合作，从相关测绘研究成果中精选了 200 余幅图纸，按照祈年殿、圜丘、皇穹宇、斋宫、神乐署五大组群以及北神厨库、南北宰牲亭、坛门等附属建筑，组织为 202 页图版，编辑成书，祈望能够裨益于世界文化遗产天坛的保护、研究与传承。

mapping of the Temple of Heaven and more than 300 drawings have been accomplished.

In this context, to better present, research and preserve the Temple of Heaven, the Administration of the Temple of Heaven Park has worked closely with the School of Architecture of Tianjin University on editing this book. Two hundreds drawings were elaborately selected in this book and were genuinely presented in two hundreds and two pages in the order of: the Complex of Hall of Prayer for Good Harvests (*Qiniandian*), the Complex of Circular Mound Altar (*Yuanqiu*), the Complex of Imperial Vault of Heaven (*Huangqiongyu*), the Complex of Palace of Abstinence (*Zhaigong*), the Complex of Divine Music Administration (*Shenyueshu*), the Northern Divine Kitchen (*Shenchu*) and Divine Storeroom (*Shenku*), the southern and northern Pavilion of Immolation (*Zaishengting*) and accessory buildings such as altar gates. As the editorial team, we really hope this book could contribute to the preservation, research and inheritance of the Temple of Heaven.

注　释

〇二　康熙朝仅在十二年，改神乐观正殿太和殿名为凝禧殿。

〇一　按康熙朝《御制律吕正义》所载，古尺长为营造尺的八寸一分，合九九天数。

Notes

① The only change was in the twelfth year of the reign of Fulin's son, the Kangxi Emperor (1763), when the main hall of the Divine Music Monastery (*Shenyue Guan*)—the Hall of Supreme Harmony (*Taihe Dian*)—changed its name to the Hall of Accumulated Happiness (*Ningxi Dian*).

② According to *Imperial Standard Interpretation of Tone-System (Yuzhi Lvlv Zhengyi)* published in the reign of the Kangxi Emperor, the length of 100 *chi* in ancient times equaled to the length of 81 chi used by engineers in public works in the Qing Dynasty. Eighty-one is the square of nine, and was regarded as the number of heaven in Chinese culture.

图版

Drawings

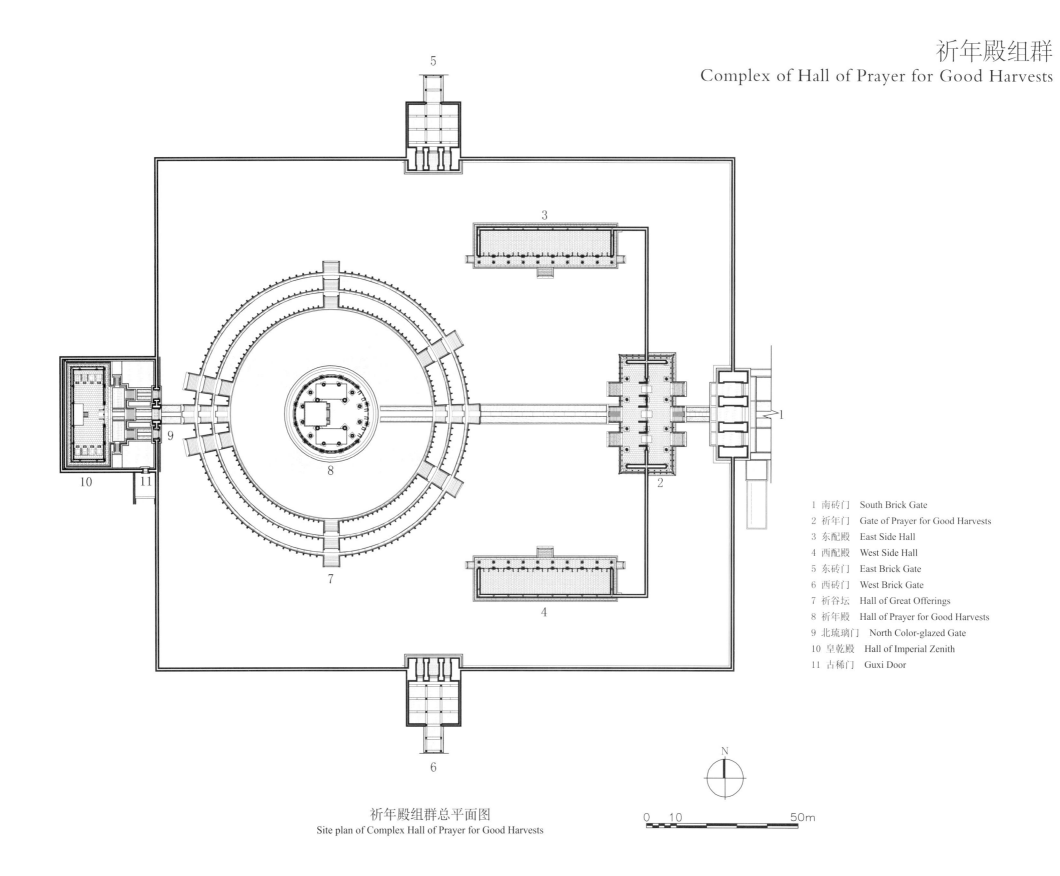

1 南砖门　South Brick Gate
2 祈年门　Gate of Prayer for Good Harvests
3 东配殿　East Side Hall
4 西配殿　West Side Hall
5 东砖门　East Brick Gate
6 西砖门　West Brick Gate
7 祈谷坛　Hall of Great Offerings
8 祈年殿　Hall of Prayer for Good Harvests
9 北琉璃门　North Color-glazed Gate
10 皇乾殿　Hall of Imperial Zenith
11 古稀门　Guxi Door

祈年殿组群总平面图
Site plan of Complex Hall of Prayer for Good Harvests

0　10　　　　50m

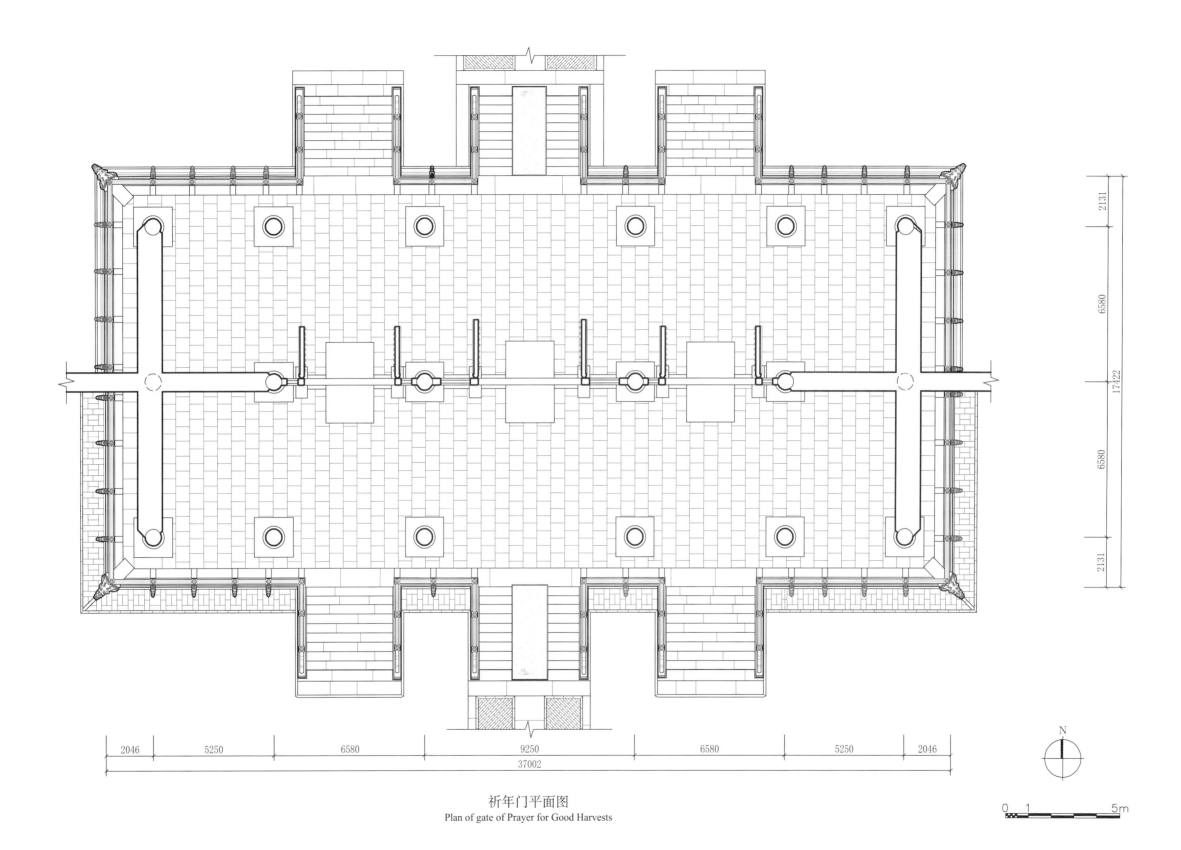

祈年门平面图
Plan of gate of Prayer for Good Harvests

0 1 5m

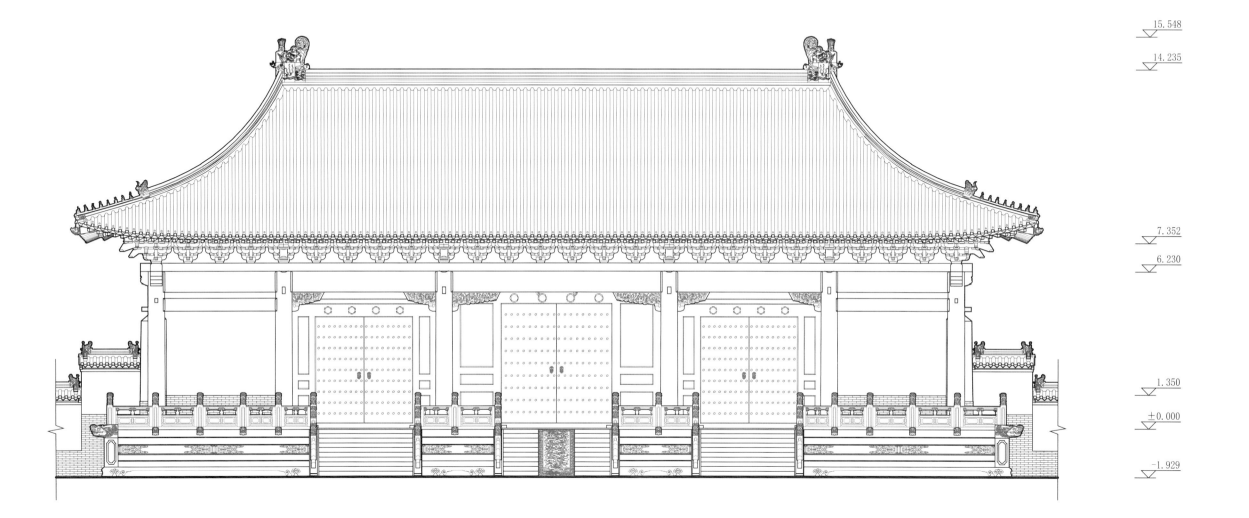

15.548

14.235

7.352

6.230

1.350

±0.000

-1.929

2046 5250 6580 9250 6580 5250 2046

37002

祈年门正立面图
Front elevation of gate of Prayer for Good Harvests

0　1　　　　5m

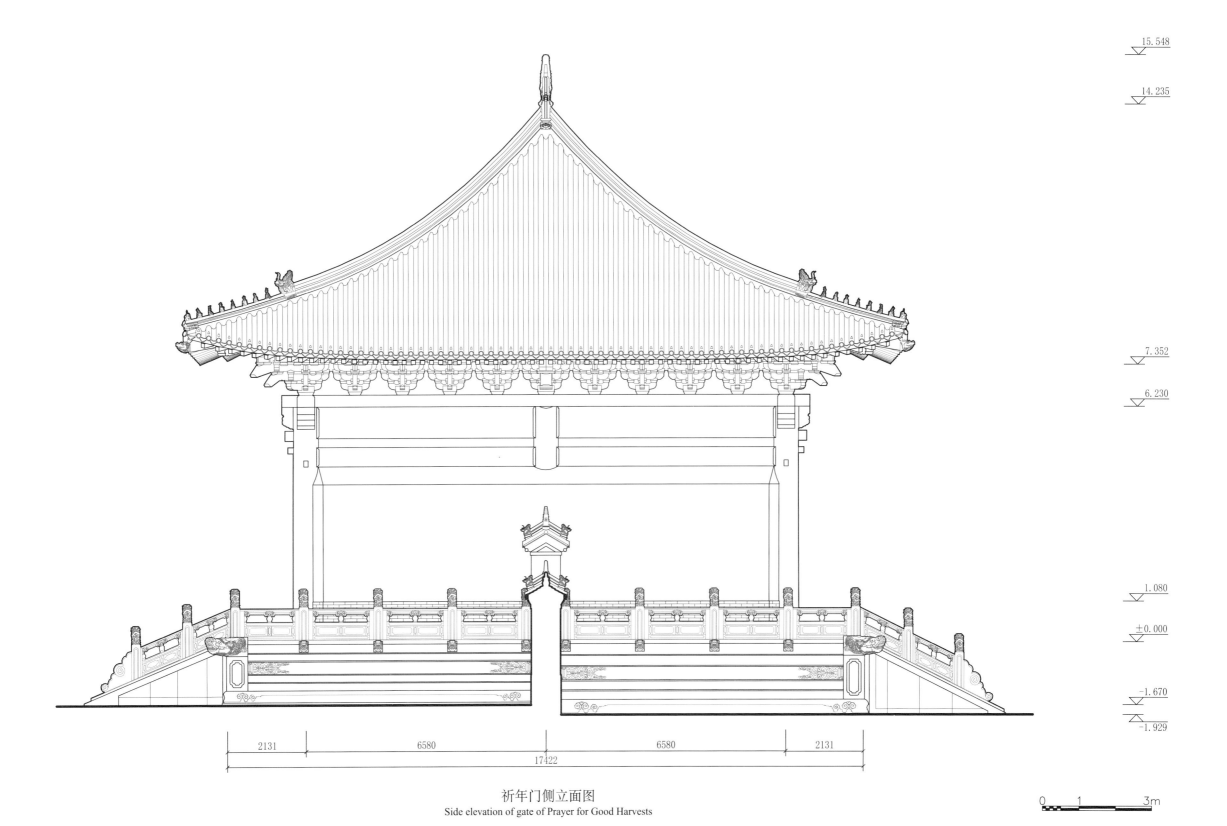

15.548

14.235

7.352

6.230

1.080

±0.000

-1.670

-1.929

2131　　　6580　　　6580　　　2131

17422

祈年门侧立面图
Side elevation of gate of Prayer for Good Harvests

0　　1　　　3m

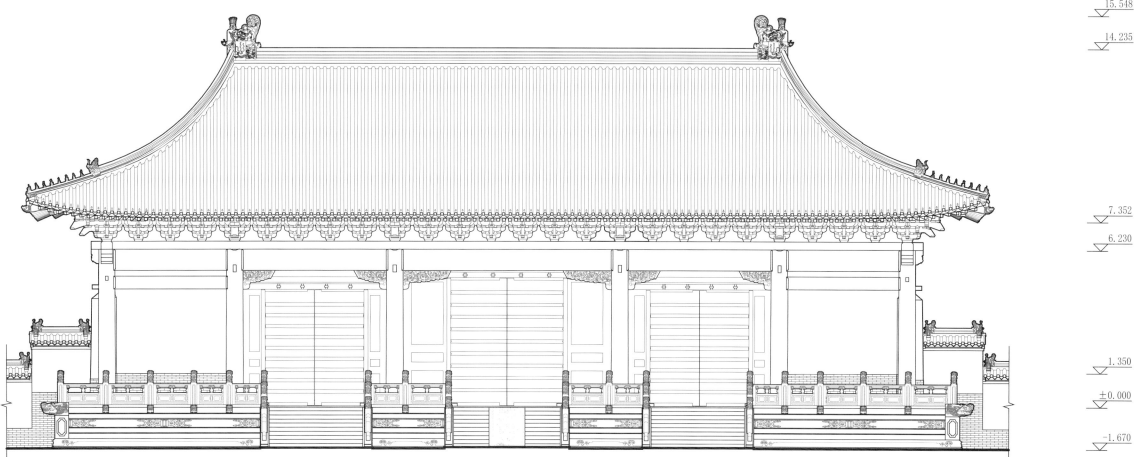

15.548

14.235

7.352

6.230

1.350

±0.000

−1.670

| 2046 | 5250 | 6580 | 9250 | 6580 | 5250 | 2046 |

37002

祈年门背立面图
Rear elevation of gate of Prayer for Good Harvests

0 1 5m

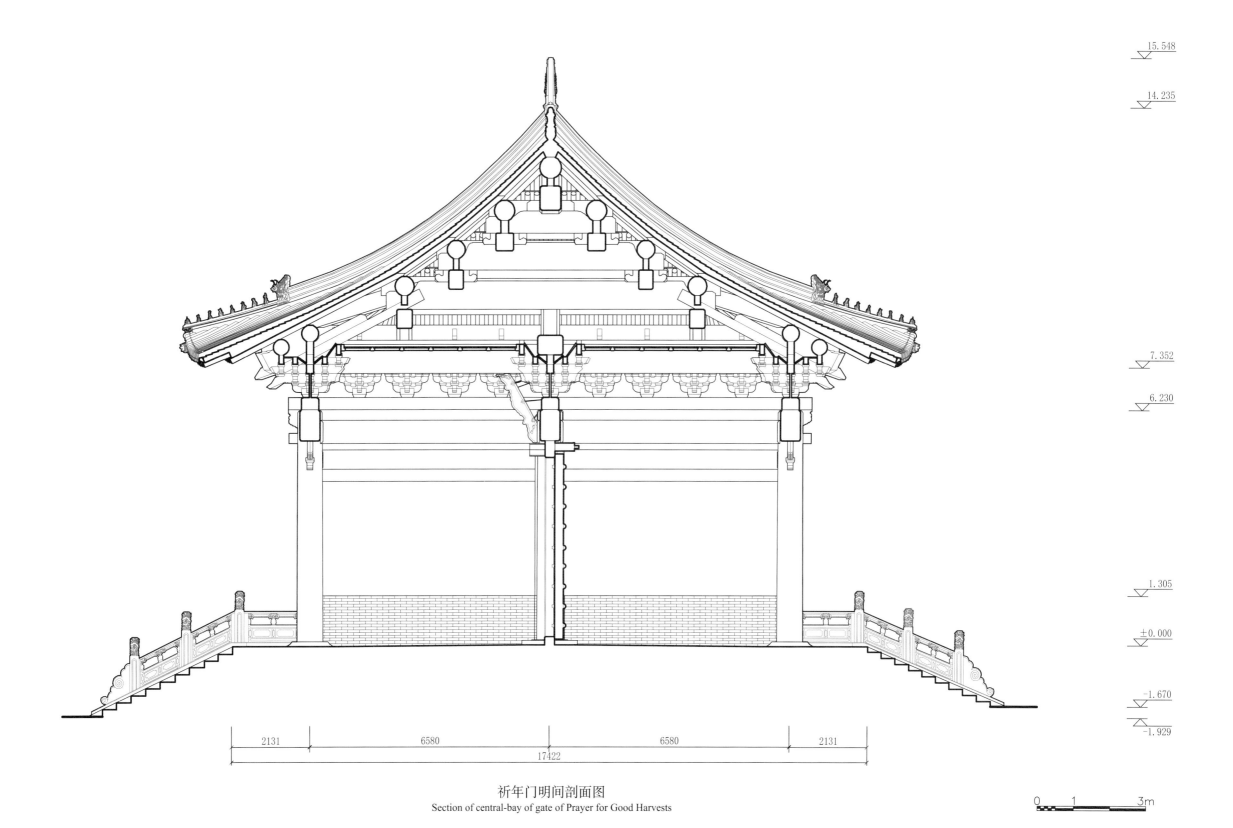

15.548

14.235

7.352

6.230

1.305

±0.000

-1.670

-1.929

2131　　6580　　6580　　2131

17422

祈年门明间剖面图
Section of central-bay of gate of Prayer for Good Harvests

0　1　3m

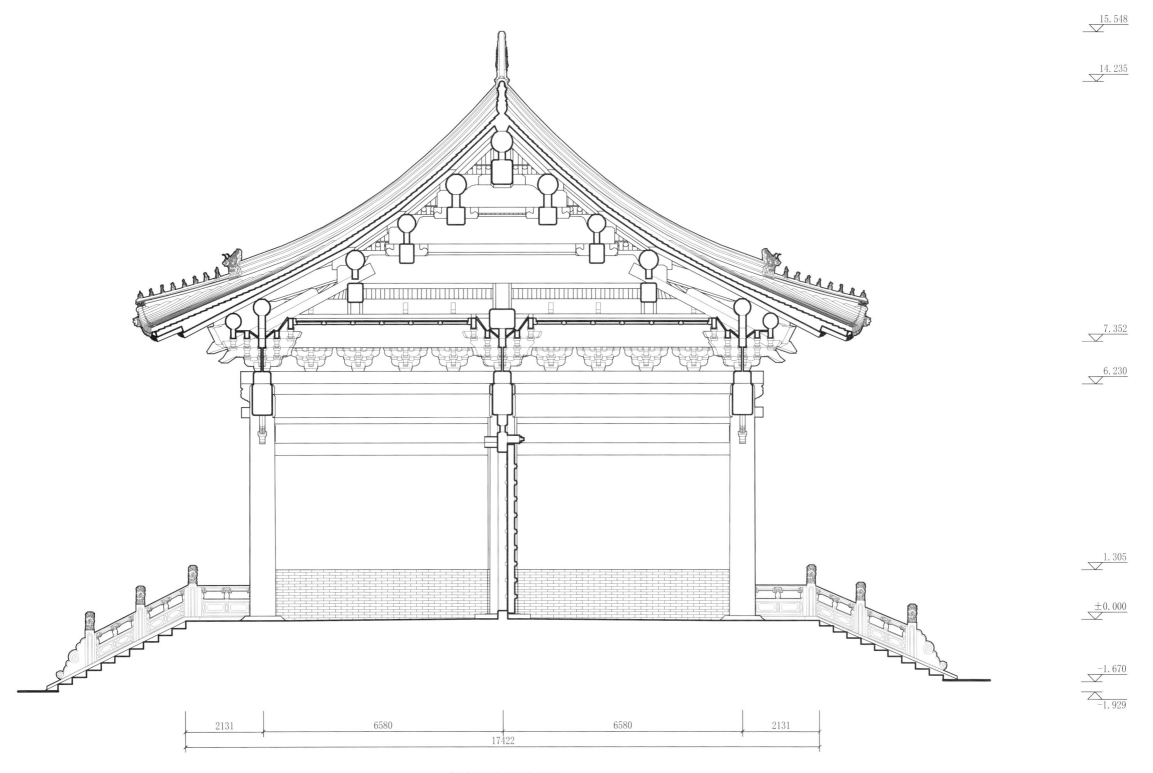

15.548

14.235

7.352

6.230

1.305

±0.000

-1.670

-1.929

2131   6580   6580   2131

17422

祈年门次间剖面图
Section of side-bay of gate of Prayer for Good Harvests

0   1   3m

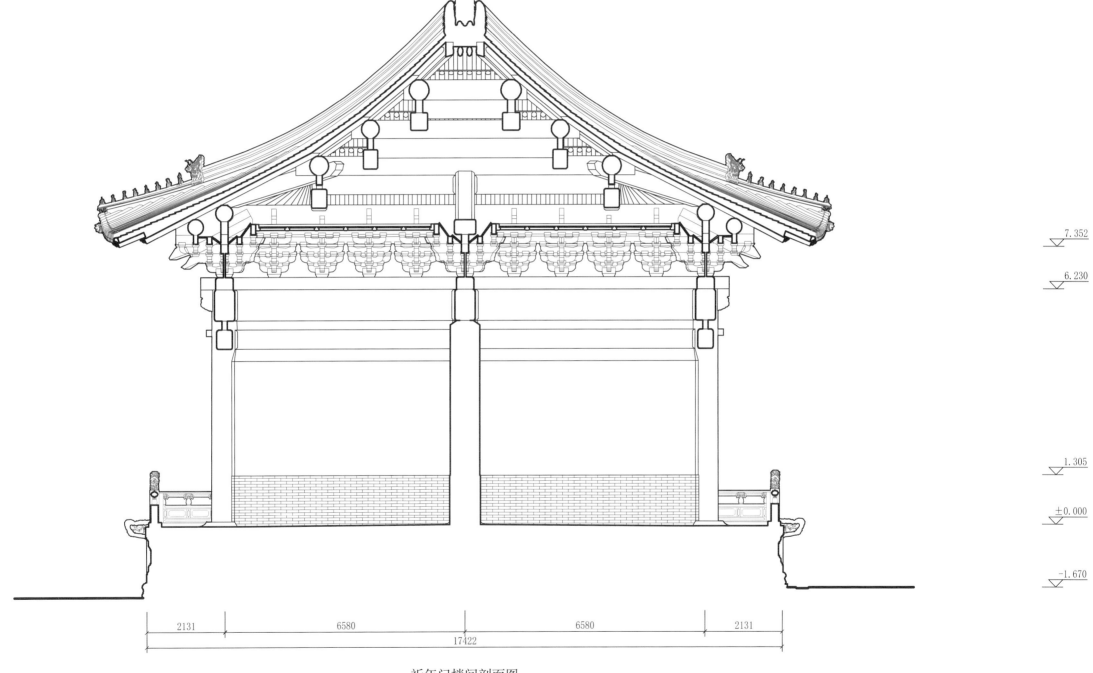

7.352

6.230

1.305

±0.000

-1.670

2131　6580　6580　2131

17422

祈年门梢间剖面图
Section of second-to-last-bay of gate of Prayer for Good Harvests

0　1　3m

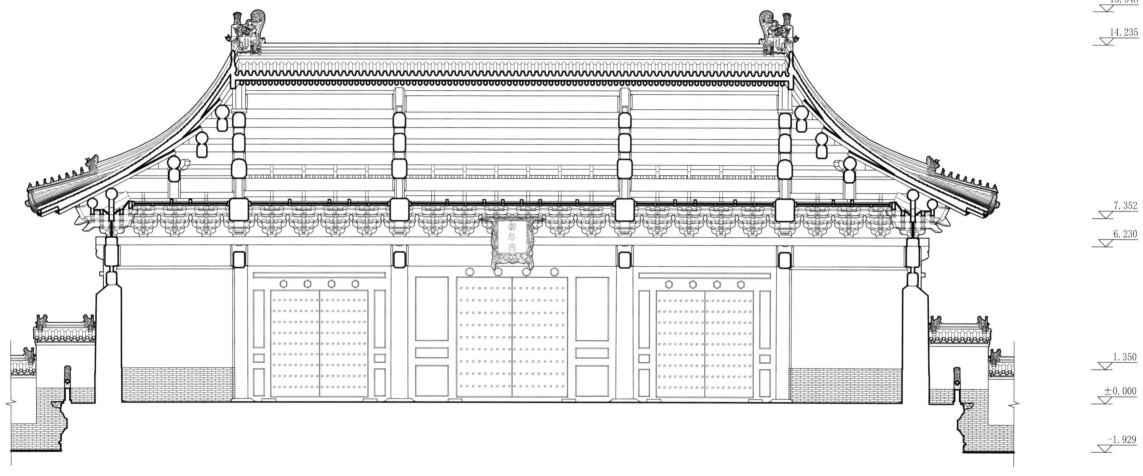

15.548

14.235

7.352

6.230

040

1.350

±0.000

-1.929

2046　5250　6580　9250　6580　5250　2046

37002

祈年门纵剖面图
Longitudinal section of gate of Prayer for Good Harvests

0　1　5m

祈年门丹陛大样图
Danbi of gate of Prayer for Good Harvests

0 0.1   0.5m

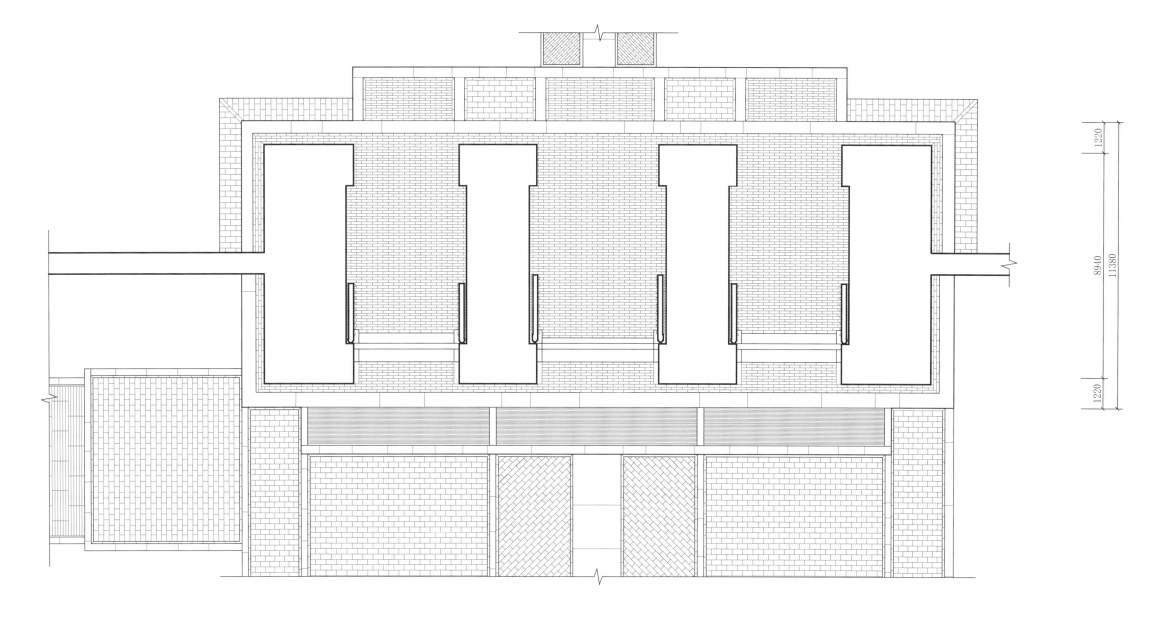

1200 1909 7429 8200 7429 1909 1200

29276

1220 8940 11380 1220

祈年殿南砖门平面图
Plan of south brick gate of Hall of Prayer for Good Harvests

N

0 1 5m

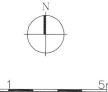

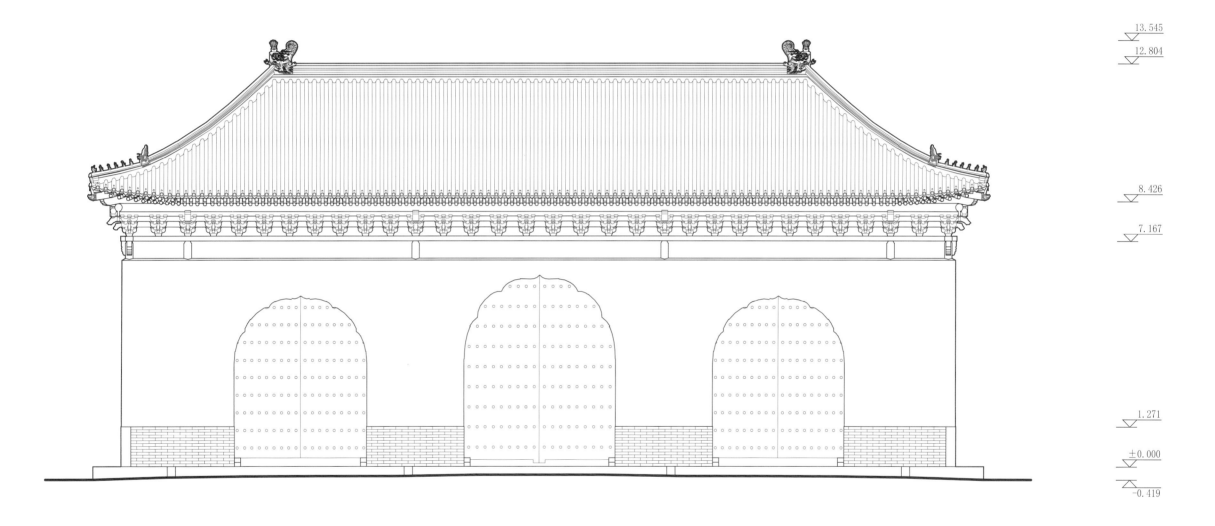

13.545

12.804

8.426

7.167

1.271

±0.000

-0.419

| 1200 | 1909 | 7429 | 8200 | 7429 | 1909 | 1200 |

29276

祈年殿南砖门正立面图
Front elevation of south brick gate of Hall of Prayer for Good Harvests

0 1 3m

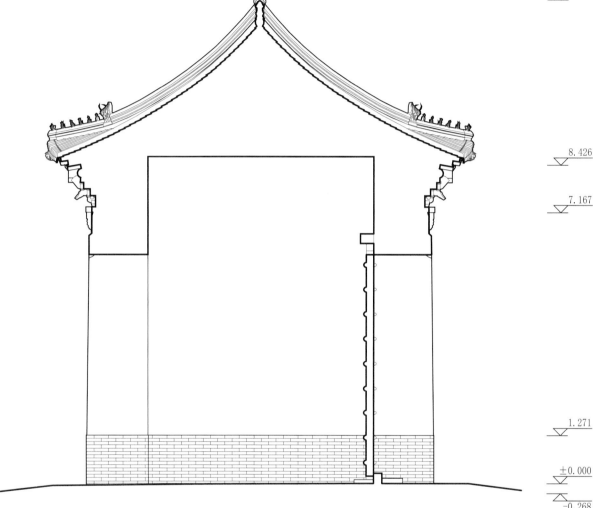

祈年殿南砖门侧立面图
Side elevation of south brick gate of Hall of Prayer for Good Harvests

祈年殿南砖门明间剖面图
Section of central-bay of south brick gate of Hall of Prayer for Good Harvests

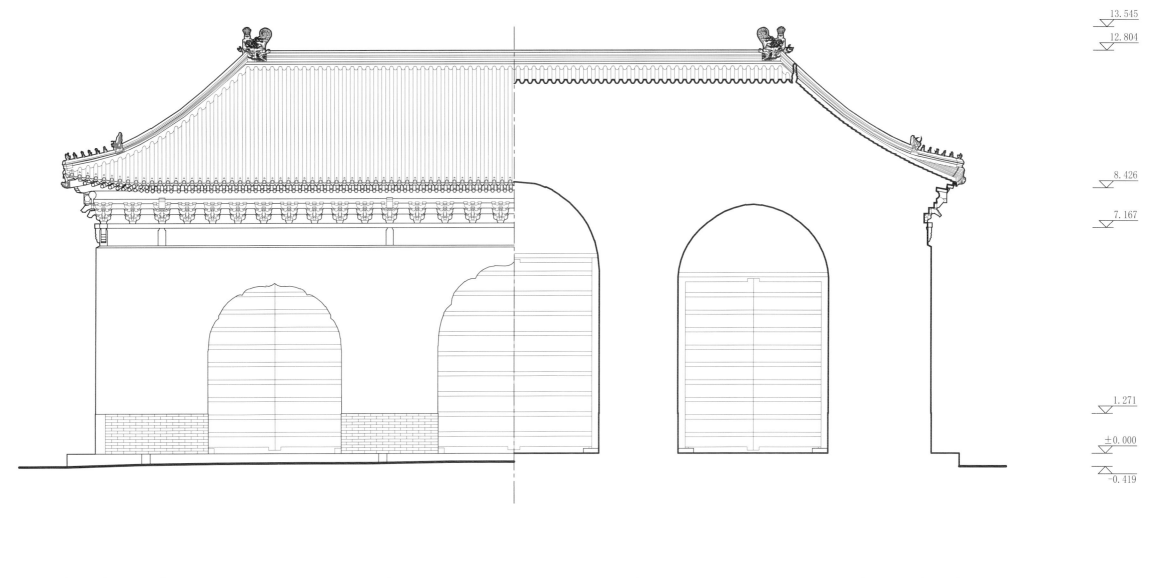

祈年殿南砖门背立面图
Rear elevation of south brick gate of Hall of Prayer for Good Harvests

祈年殿南砖门纵剖面图
Longitudinal section of south brick gate of Hall of Prayer for Good Harvests

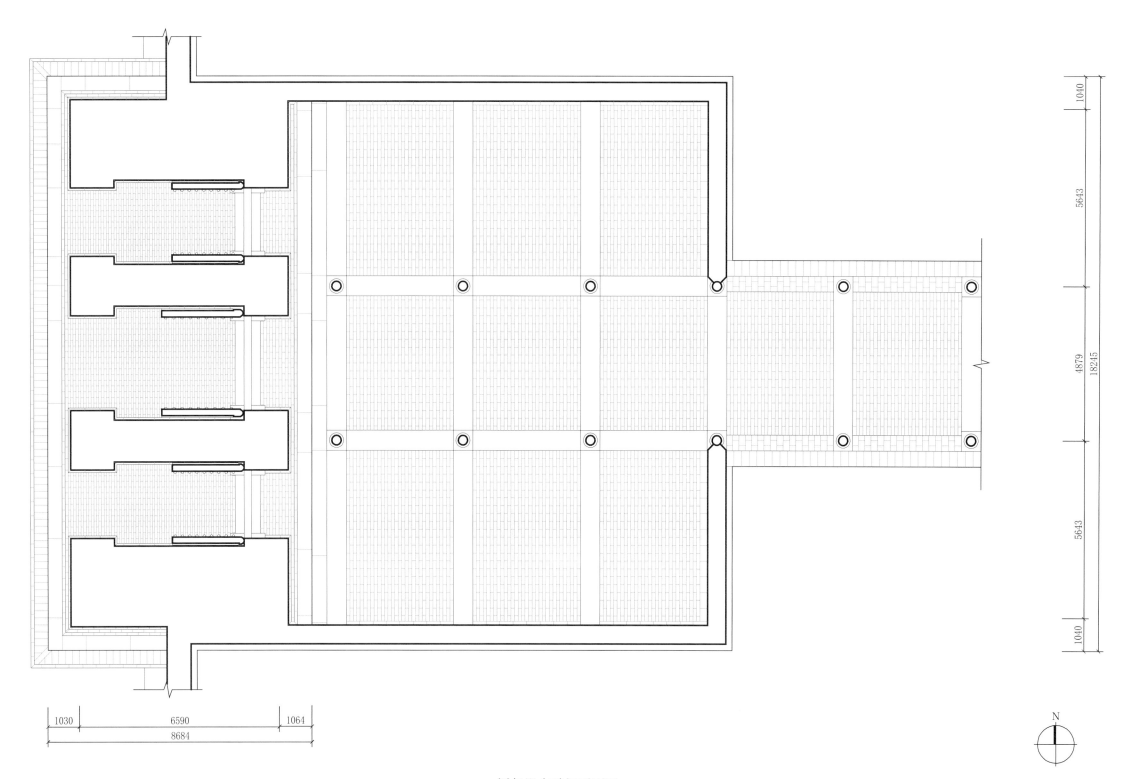

祈年殿东砖门平面图
Plan of east brick gate of Hall of Prayer for Good Harvests

0 1 3m

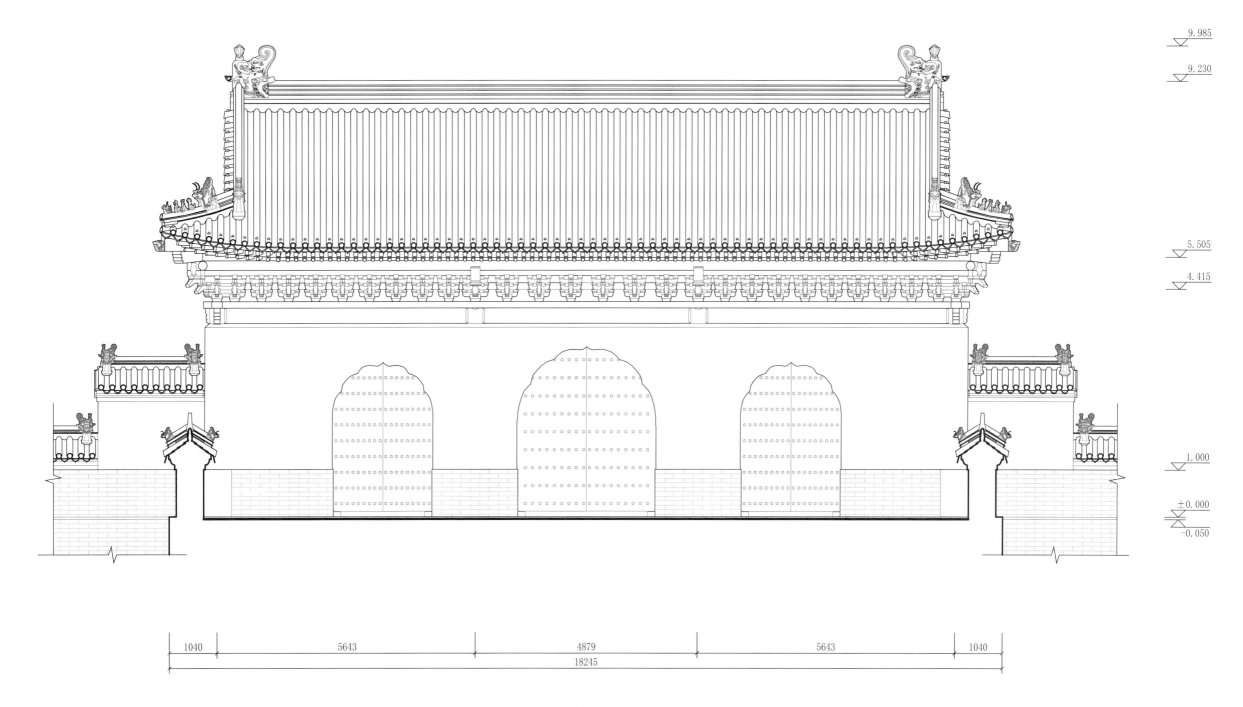

9.985

9.230

5.505

4.415

1.000

±0.000

-0.050

1040	5643	4879	5643	1040

18245

祈年殿东砖门正立面图
Front elevation of east brick gate of Hall of Prayer for Good Harvests

0 1 2m

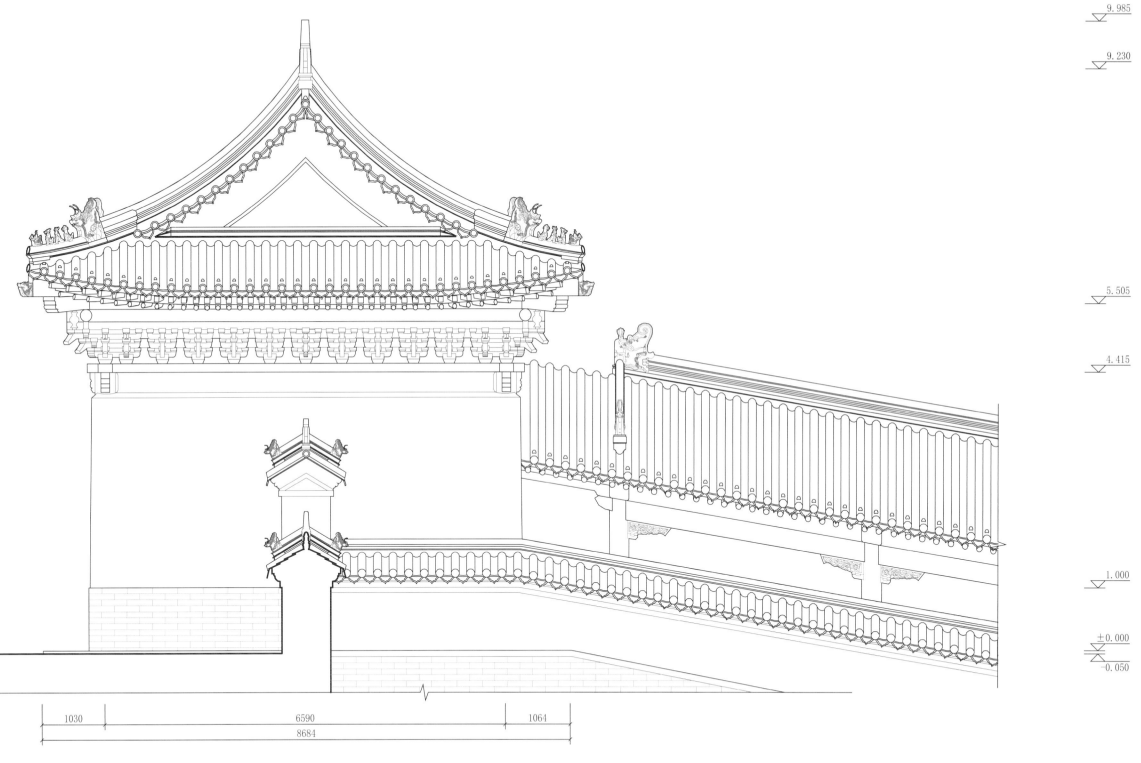

9.985

9.230

5.505

4.415

1.000

±0.000

-0.050

1030   6590   1064

8684

祈年殿东砖门侧立面图
Side elevation of east brick gate of Hall of Prayer for Good Harvests

0   1   2m

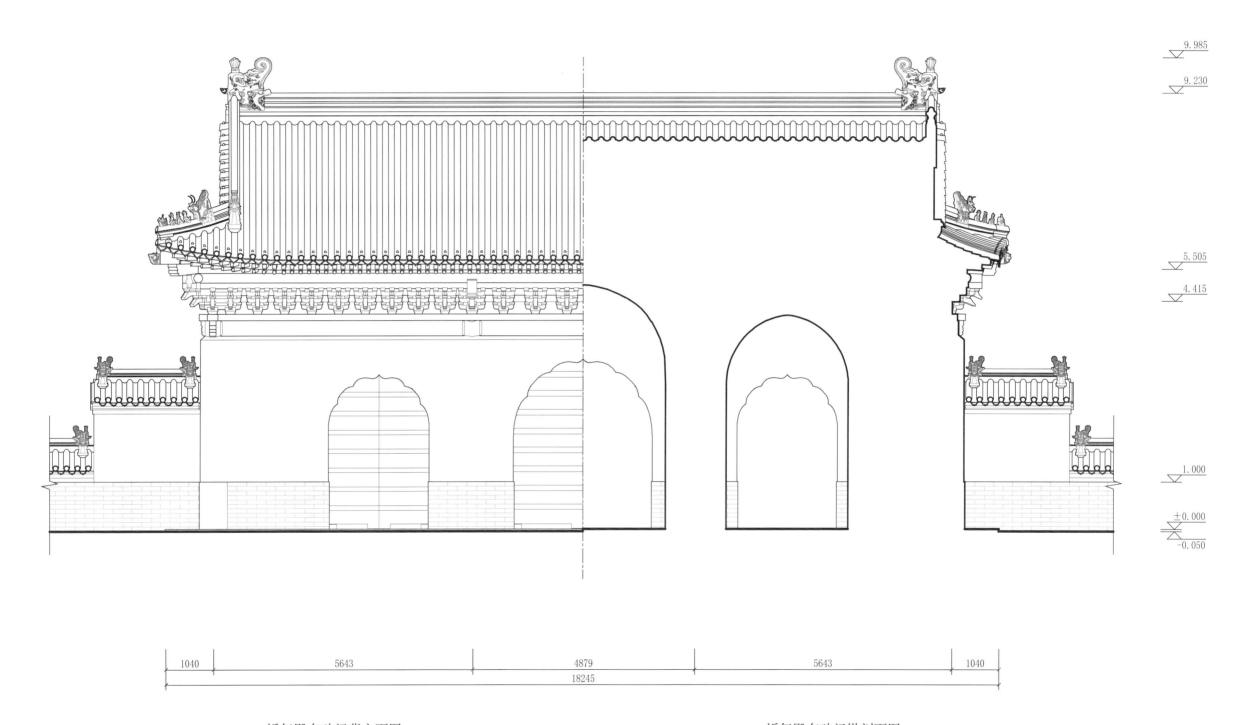

9.985
9.230
5.505
4.415
1.000
±0.000
-0.050

1040　5643　4879　5643　1040
18245

祈年殿东砖门背立面图
Rear elevation of east brick gate of Hall of Prayer for Good Harvests

祈年殿东砖门纵剖面图
Longitudinal elevation of east brick gate of Hall of Prayer for Good Harvests

0　1　2m

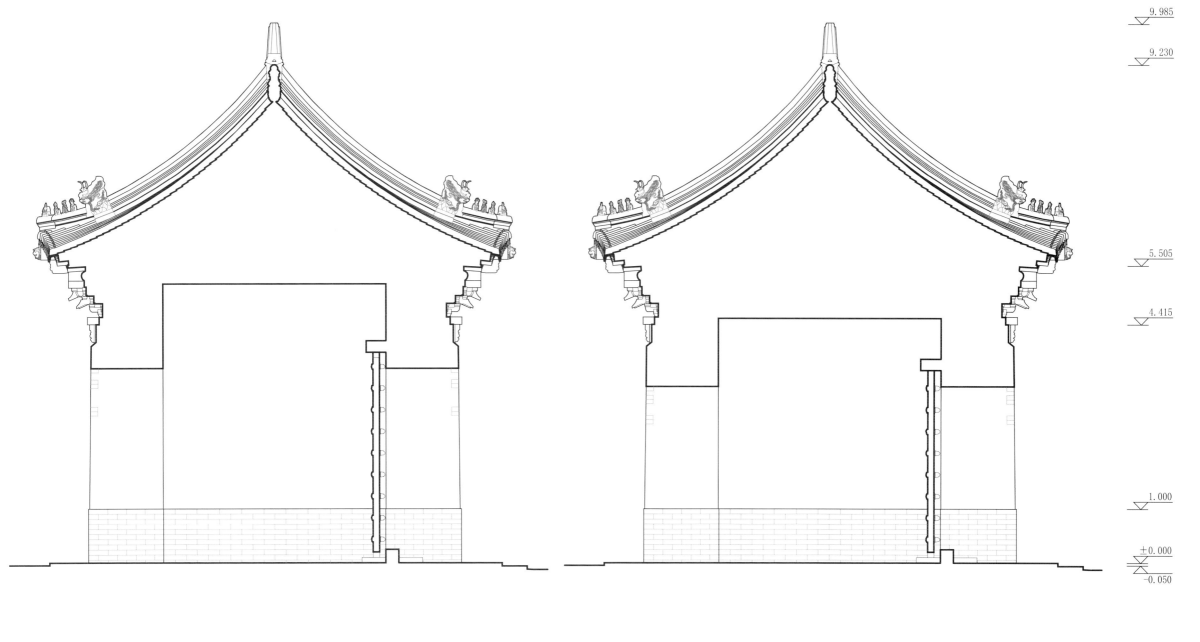

9.985

9.230

5.505

4.415

1.000

±0.000

-0.050

| 1030 | 6590 | 1064 |

8684

| 1030 | 6590 | 1064 |

8684

祈年殿东砖门明间剖面图
Section of central-bay of east brick gate of Hall of Prayer for Good Harvests

祈年殿东砖门次间剖面图
Section of side-bay of east brick gate of Hall of Prayer for Good Harvests

0   1   2m

祈年殿及祈谷坛平面图
Plan of Hall of Prayer for Good Harvests and Hall of Great Offerings

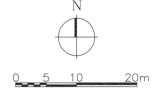

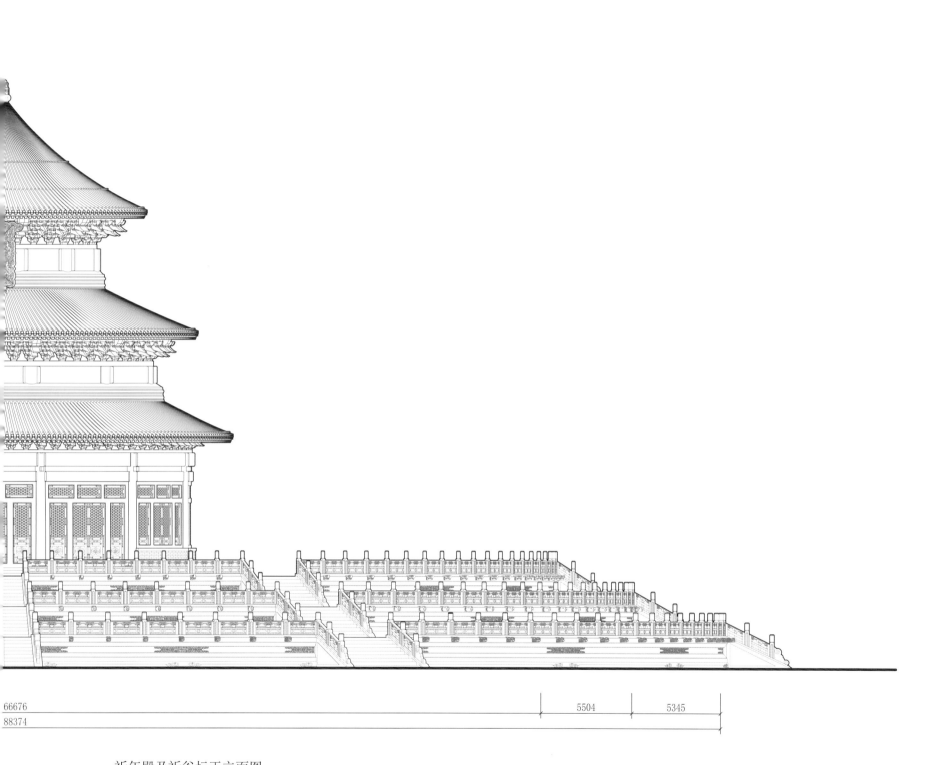

31.778

27.104

20.330

18.564

13.283

11.729

7.287
6.700

±0.000

-2.260

-4.040

-5.820

66676

88374

5504 　 5345

祈年殿及祈谷坛正立面图
Front elevation of Hall of Prayer for Good Harvests and Hall of Great Offerings

0　1　　　　　5

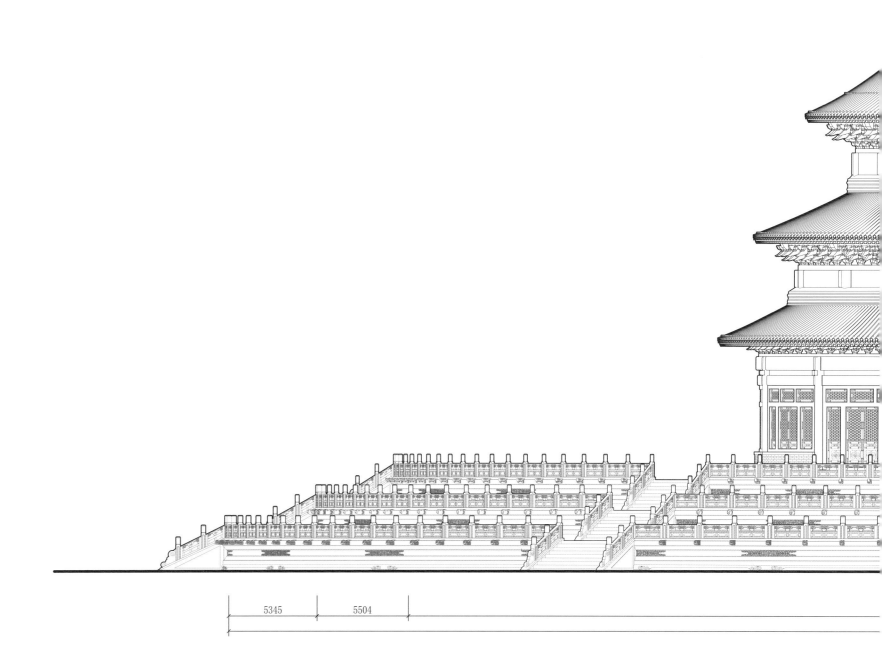

5345   5504

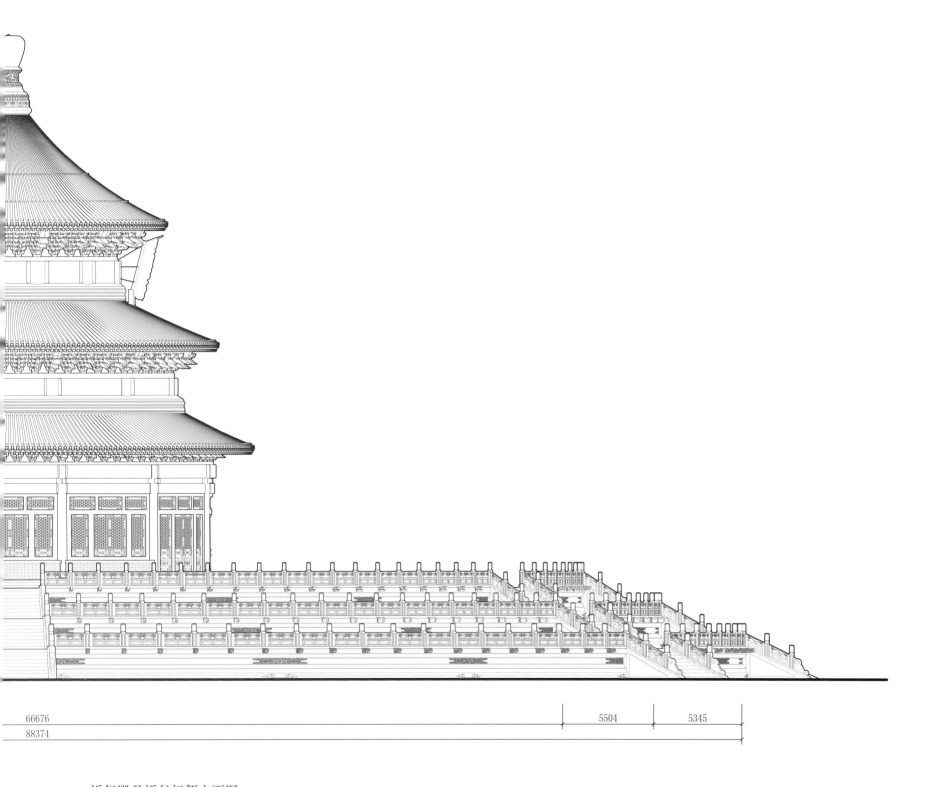

31.778

27.104

20.330

18.564

13.283

11.729

7.287
6.700

±0.000

-2.260

-4.040

-5.820

66676

88374

5504   5345

祈年殿及祈谷坛侧立面图

Side elevation of Hall of Prayer for Good Harvests and Hall of Great Offerings

0 1   5m

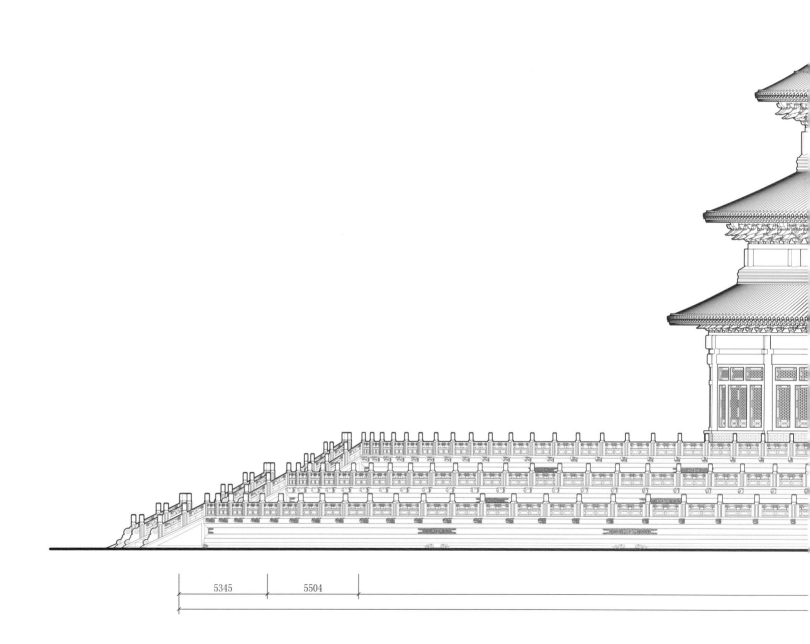

5345   5504

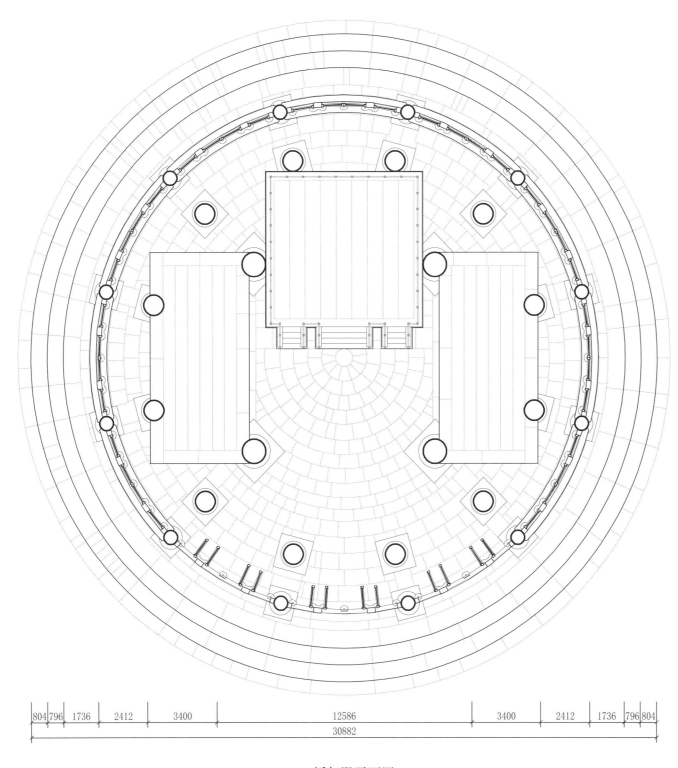

| 804 | 796 | 1736 | 2412 | 3400 | 12586 | 3400 | 2412 | 1736 | 796 | 804 |

30882

祈年殿平面图
Plan of Hall of Prayer for Good Harvests

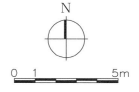

N

0 1   5m

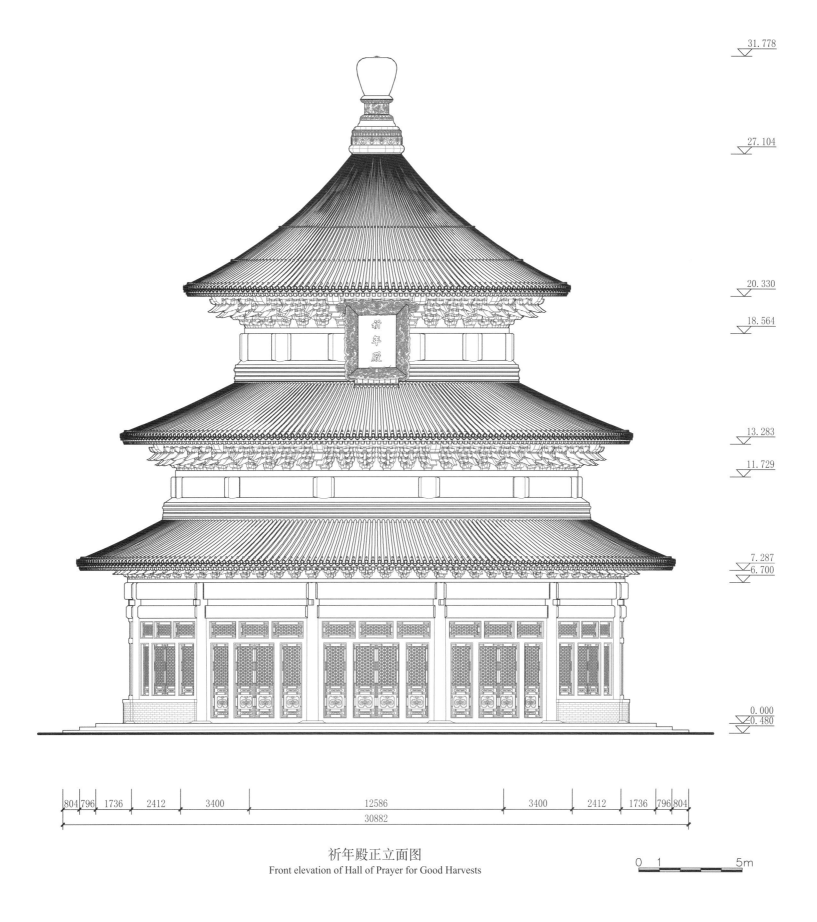

31.778

27.104

20.330

18.564

13.283

11.729

7.287
6.700

0.000
0.480

804 796 1736 2412 3400 12586 3400 2412 1736 796 804
30882

祈年殿正立面图
Front elevation of Hall of Prayer for Good Harvests

0 1 5m

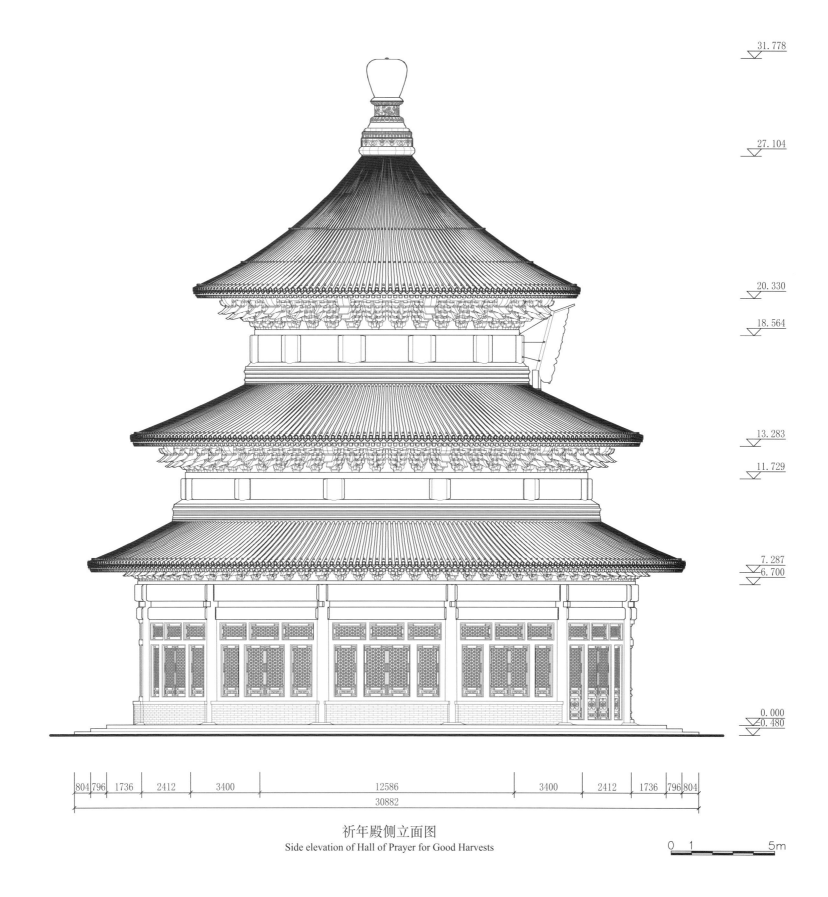

31.778

27.104

20.330

18.564

13.283

11.729

7.287
6.700

0.000
-0.480

804 796 1736 2412 3400 12586 3400 2412 1736 796 804

30882

祈年殿侧立面图
Side elevation of Hall of Prayer for Good Harvests

0 1 5m

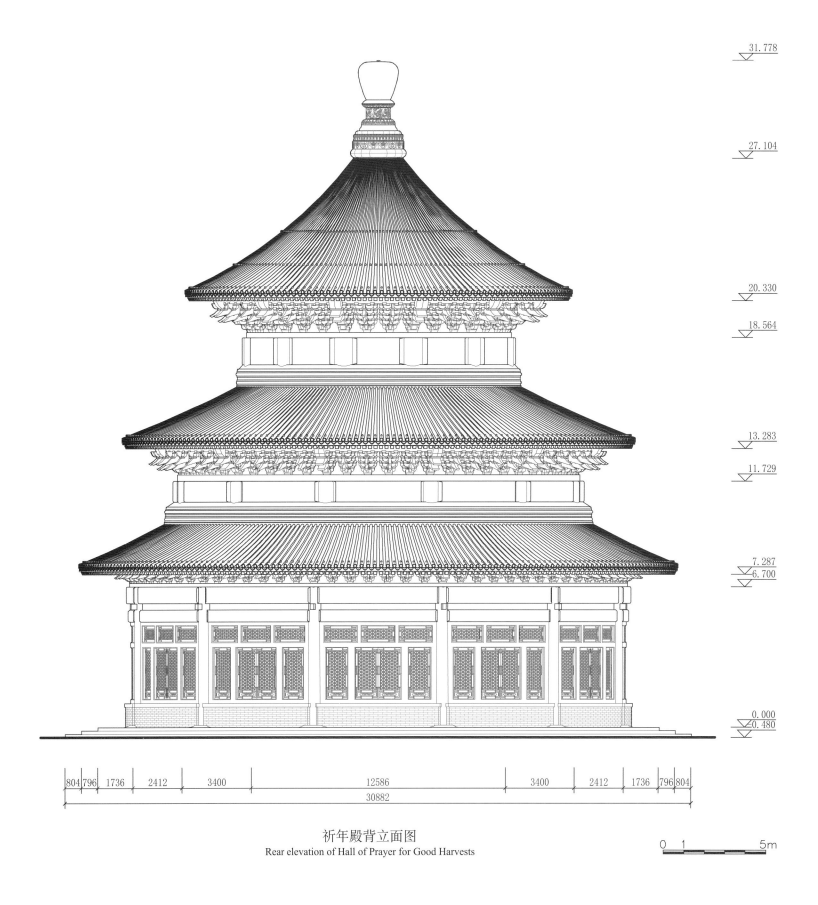

31.778

27.104

20.330

18.564

13.283

11.729

7.287
6.700

0.000
-0.480

804 796 1736 2412 3400 12586 3400 2412 1736 796 804
30882

祈年殿背立面图
Rear elevation of Hall of Prayer for Good Harvests

0 1 5m

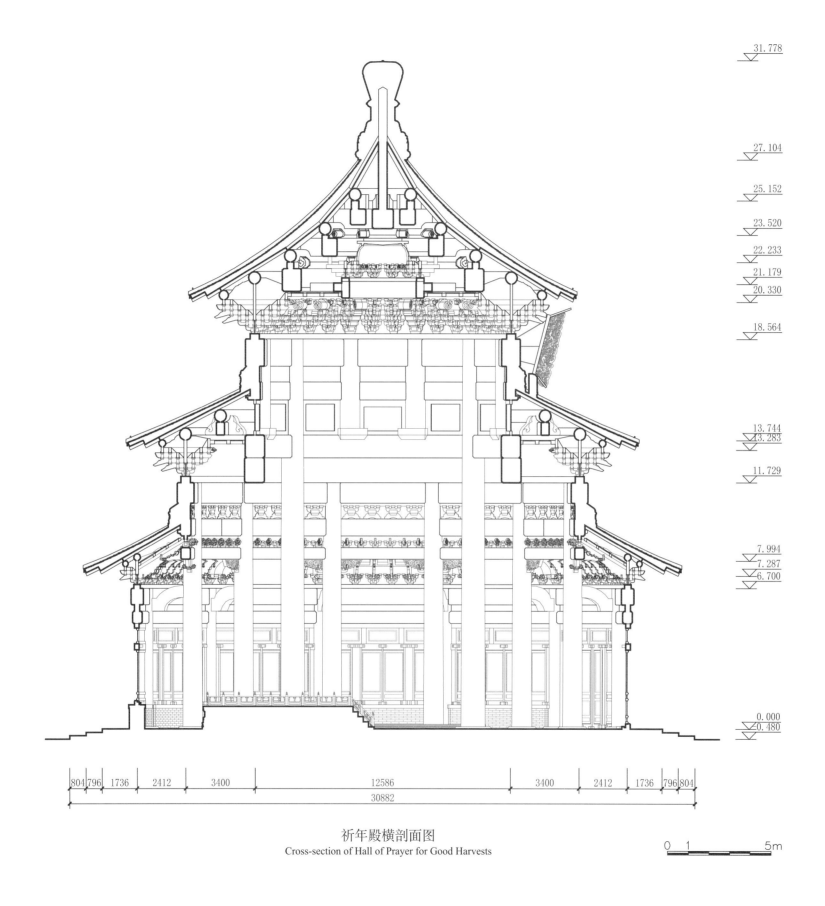

31.778

27.104

25.152

23.520

22.233

21.179

20.330

18.564

13.744
13.283

11.729

7.994
7.287
6.700

0.000
0.480

804 796 1736 2412 3400 12586 3400 2412 1736 796 804

30882

祈年殿横剖面图
Cross-section of Hall of Prayer for Good Harvests

0 1 5m

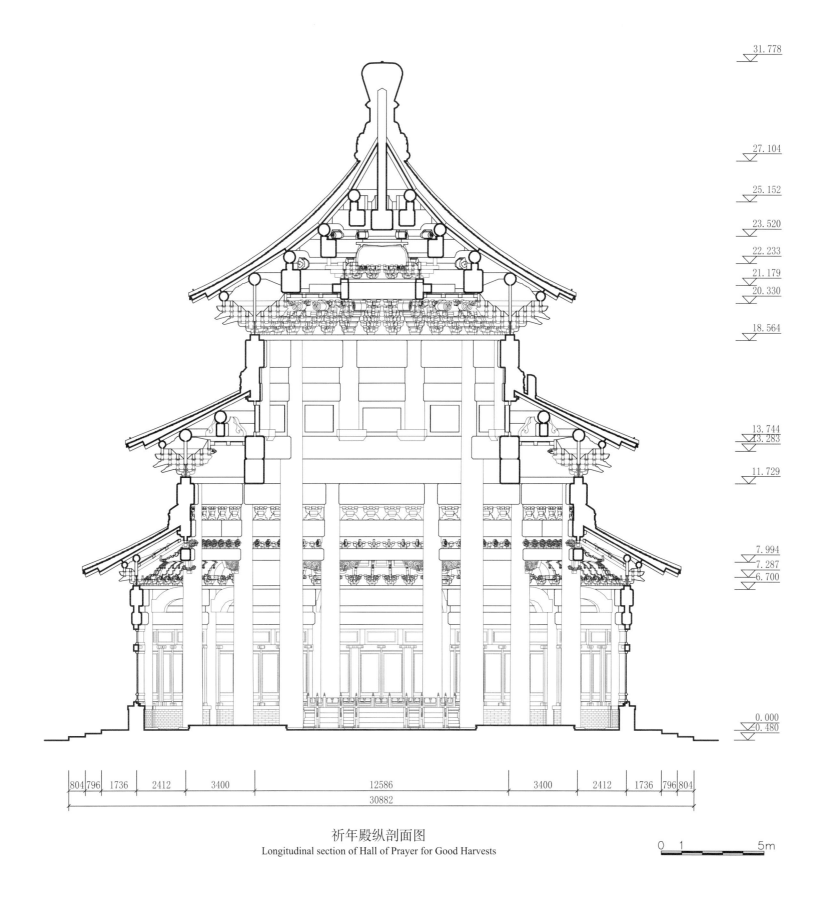

31.778

27.104

25.152

23.520

22.233

21.179

20.330

18.564

13.744
13.283

11.729

7.994
7.287
6.700

0.000
-0.480

| 804 | 796 | 1736 | 2412 | 3400 | 12586 | 3400 | 2412 | 1736 | 796 | 804 |

30882

祈年殿纵剖面图
Longitudinal section of Hall of Prayer for Good Harvests

0 1   5m

| 721 | 1750 | 2419 | 3399.5 | 4272 | 1071 | 1900 | 1071 | 4272 | 3399.5 | 2419 | 1750 | 721 |

29165

祈年殿一重檐梁架仰视图

Plan of framework of bottom double-eaves of Hall of Prayer for Good Harvests as seen from below

0 1 5m

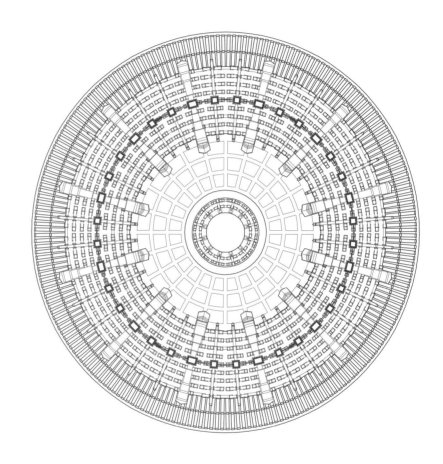

祈年殿二重檐梁架仰视图

Plan of framework of middle double-eaves of Hall of Prayer for Good Harvests as seen from below

祈年殿三重檐梁架仰视图

Plan of framework of top double-eaves of Hall of Prayer for Good Harvests as seen from below

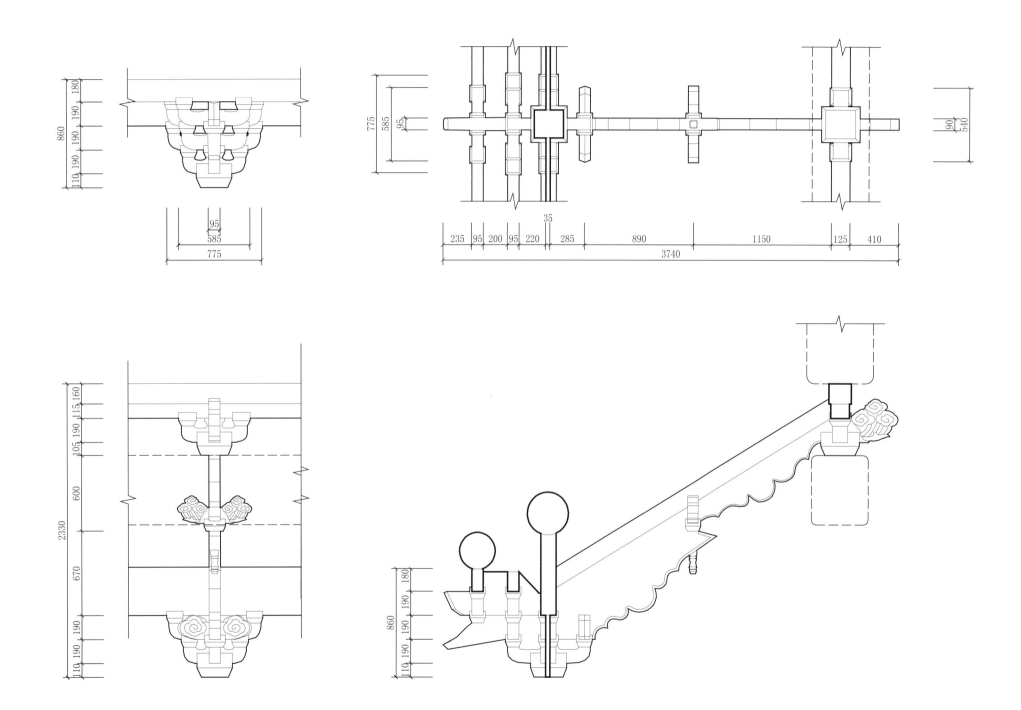

祈年殿一重檐平身科斗拱大样图
First layer of *pingshenke* bracket set of Hall of Prayer for Good Harvests

0   0.5   1m

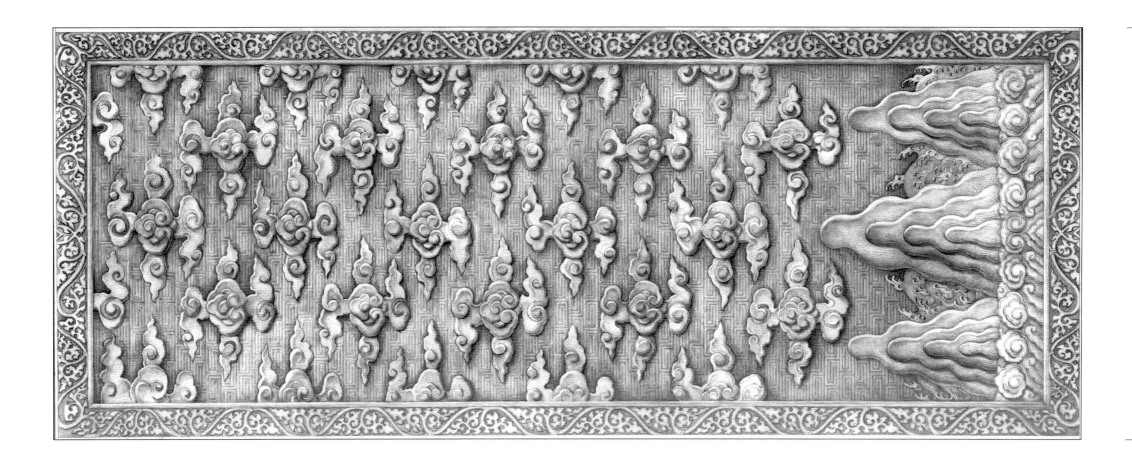

祈谷坛一层丹陛大样图
First level of *danbi* of Altar of Prayer for Good Harvests

0 0.1 0.5m

4392

1745

祈谷坛二层丹陛大样图
Second level of *danbi* of Altar of Prayer for Good Harvests

0 0.1   0.5m

4378

1745

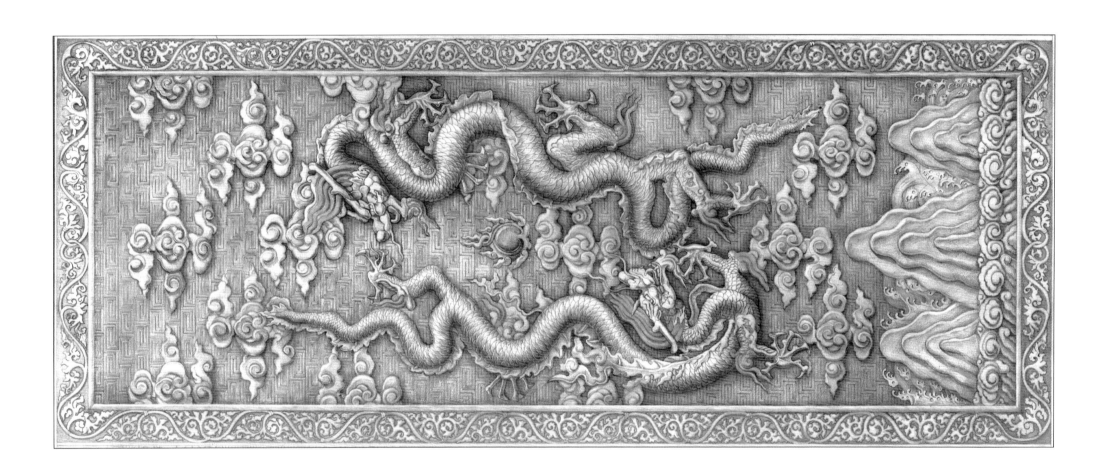

祈谷坛三层丹陛大样图
Third level of *danbi* of Altar of Prayer for Good Harvests

0   0.1   0.5m

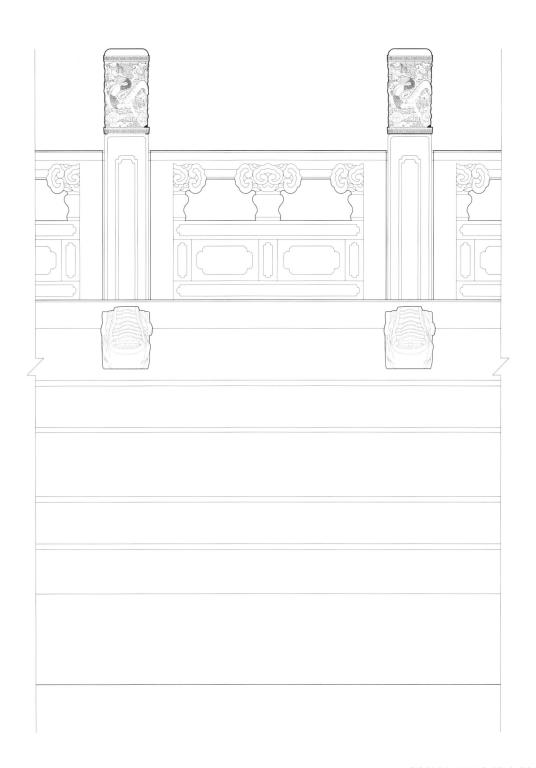
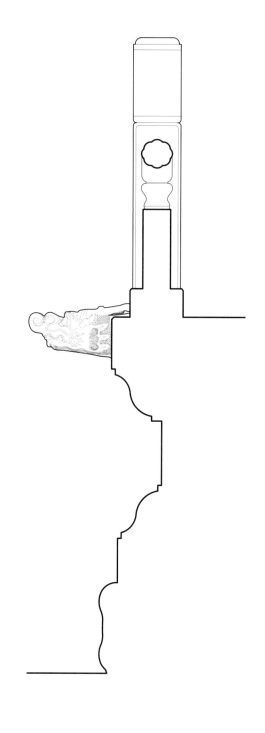

祈谷坛三层台基大样图
Third level of *taiji* of Hall of Great Offerings

0 0.1 0.5m

祈谷坛望柱柱头立面图
Elevation of *wangzhu* capital of Altar of Prayer for Good Harvests

祈谷坛望柱柱头顶视图
Plan of *wangzhu* capital of Altar of Prayer for Good Harvests

0   0.1   0.2m

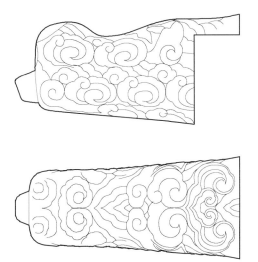
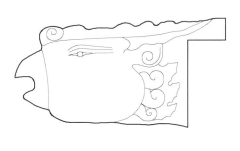
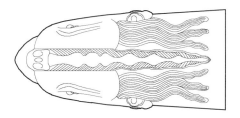
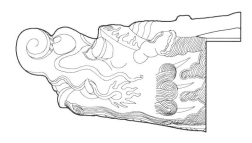
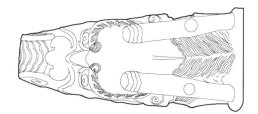

祈谷坛勾嘴大样图
Gouzui of Altar of Prayer for Good Harvests

0   0.1   0.3m

祈谷坛一层望柱柱头立面展开图
Unfolded elevation of *wangzhu* capital on first level of Altar of Prayer for Good Harvests

0   0.05   0.1m

祈谷坛二层望柱柱头立面展开图
Unfolded elevation of *wangzhu* capital on second level of Altar of Prayer for Good Harvests

0   0.05   0.1m

祈谷坛三层望柱柱头立面展开图
Unfolded elevation of *wangzhu* capital on third level of Altar of Prayer for Good Harvests

0　　0.05　　0.1m

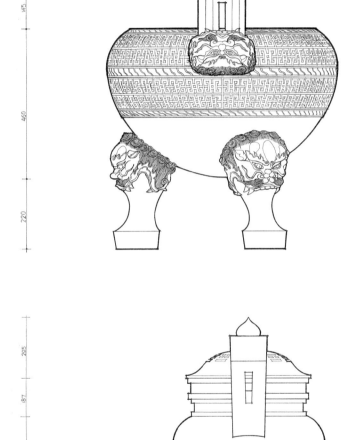
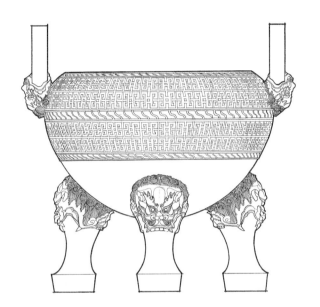
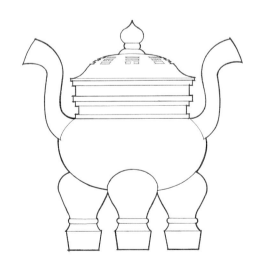
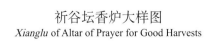
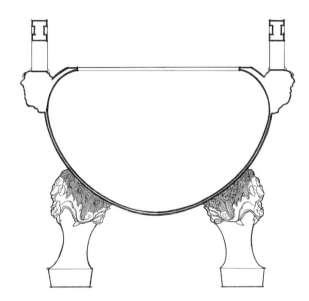
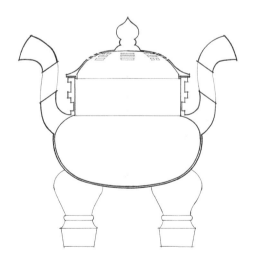

祈谷坛香炉大样图
Xianglu of Altar of Prayer for Good Harvests

0   0.1   0.3m

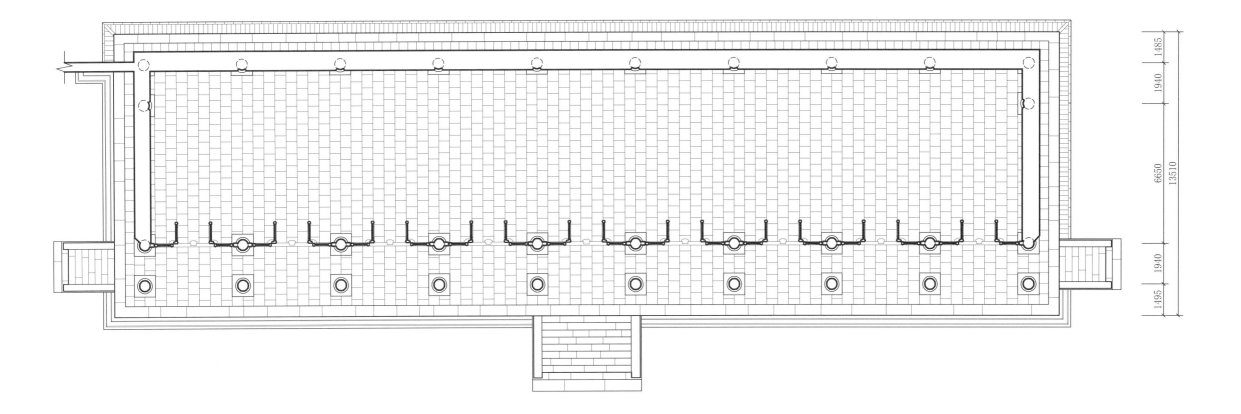

祈年殿西配殿平面图
Plan of west side hall of Hall of Prayer for Good Harvests

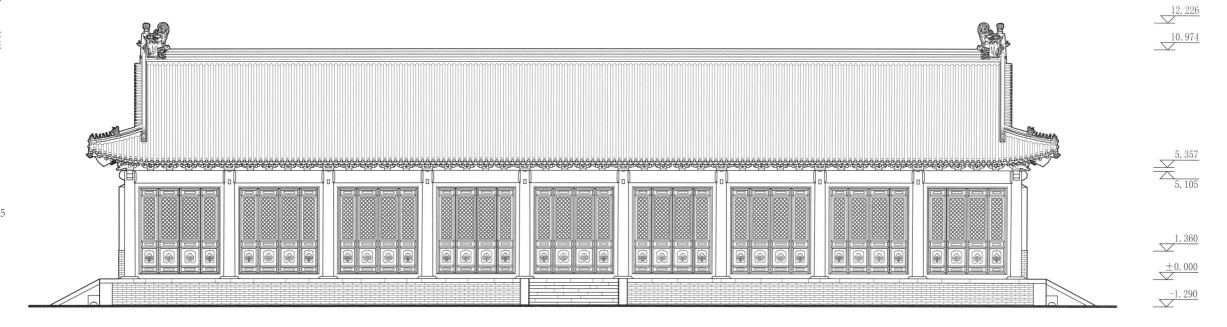

12.226

10.974

5.357

5.105

1.360

±0.000

-1.290

| 1485 | 4840 | 4840 | 4840 | 4840 | 4840 | 4840 | 4840 | 4840 | 4840 | 1485 |

46530

祈年殿西配殿正立面图

Front elevation of west side hall of Hall of Prayer for Good Harvests

0   1   5m

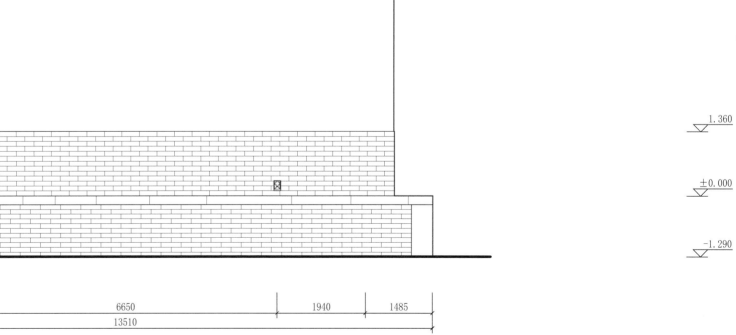

12.226

10.974

5.357
5.105

1.360

±0.000

-1.290

| 1495 | 1940 | 6650 | 1940 | 1485 |

13510

祈年殿西配殿侧立面图
Side elevation of west side hall of Hall of Prayer for Good Harvests

0   1   2m

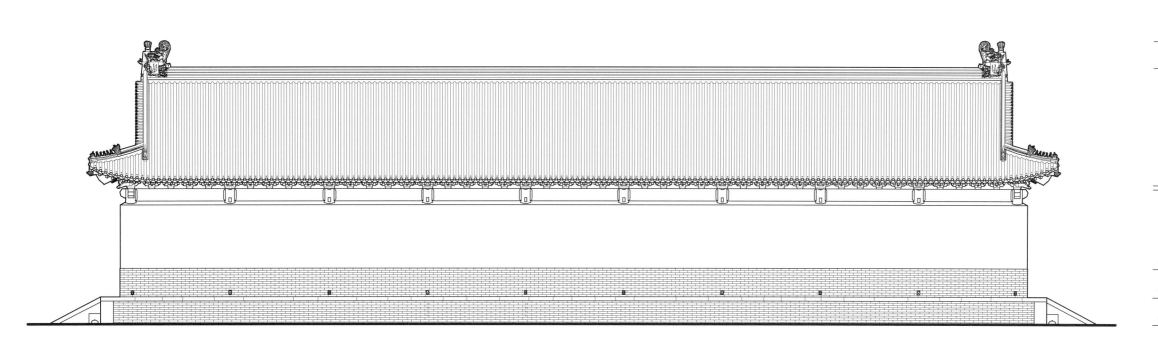

祈年殿西配殿背立面图
Rear elevation of west side hall of Hall of Prayer for Good Harvests

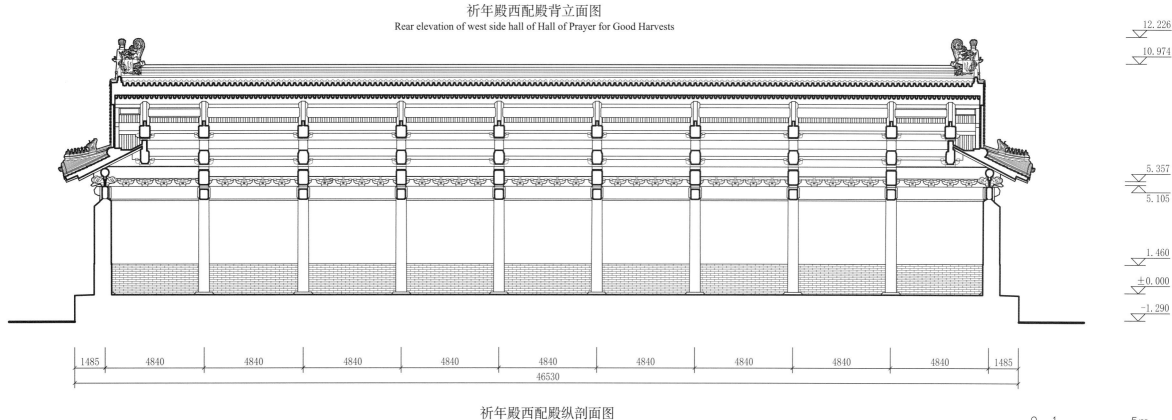

祈年殿西配殿纵剖面图
Longitudinal section of west side hall of Hall of Prayer for Good Harvests

0 1 5m

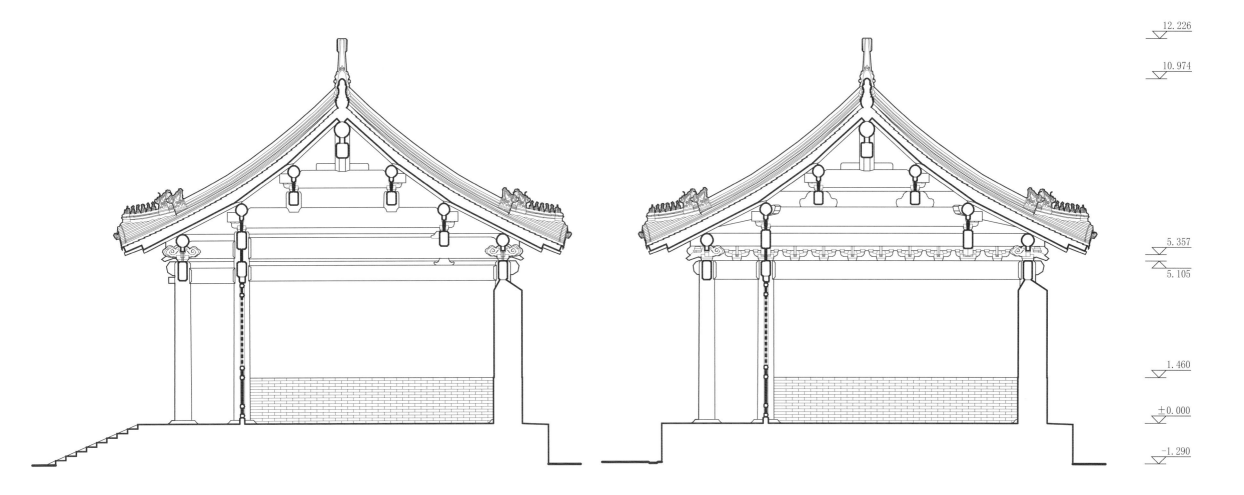

12.226

10.974

5.357

5.105

1.460

±0.000

-1.290

| 1495 | 1940 | 6650 | 1940 | 1485 |

13510

| 1495 | 1940 | 6650 | 1940 | 1485 |

13510

祈年殿西配殿明间剖面图
Section of central-bay of west side hall of Hall of Prayer for Good Harvests

祈年殿西配殿梢间剖面图
Section of second-to-last-bay of west side hall of Hall of Prayer for Good Harvests

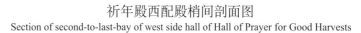

0   1   3m

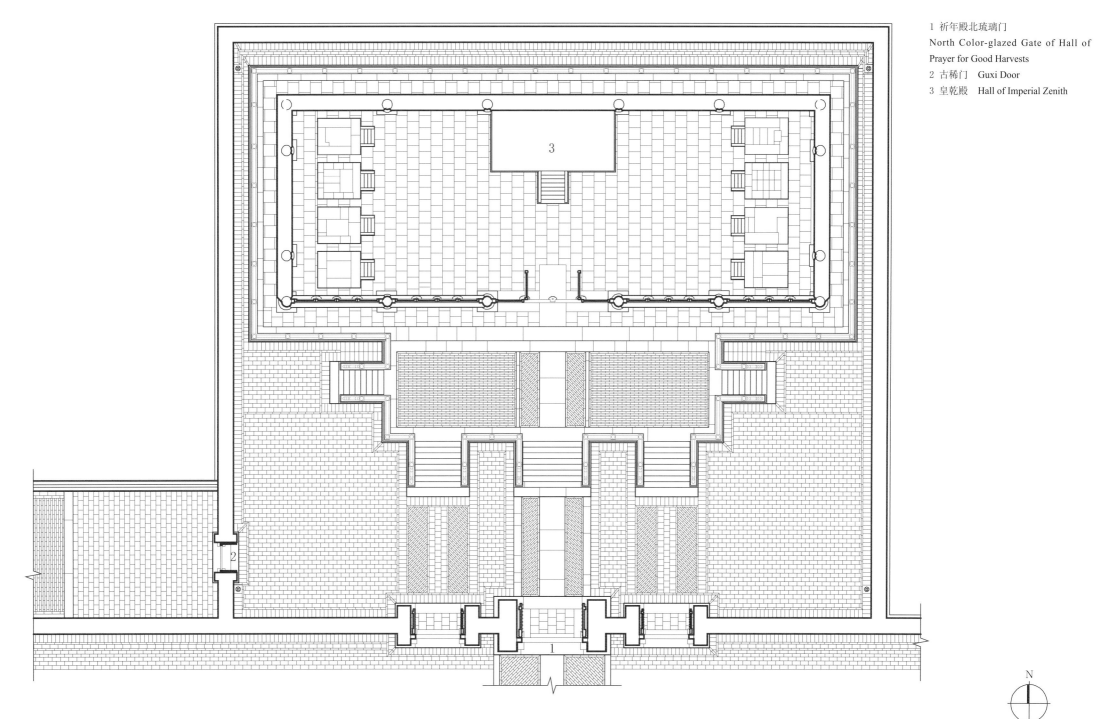

1 祈年殿北琉璃门
North Color-glazed Gate of Hall of
Prayer for Good Harvests
2 古稀门　Guxi Door
3 皇乾殿　Hall of Imperial Zenith

皇乾殿院落总平面图
Site plan of courtyard of Hall of Imperial Zenith

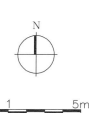

0 1   5m

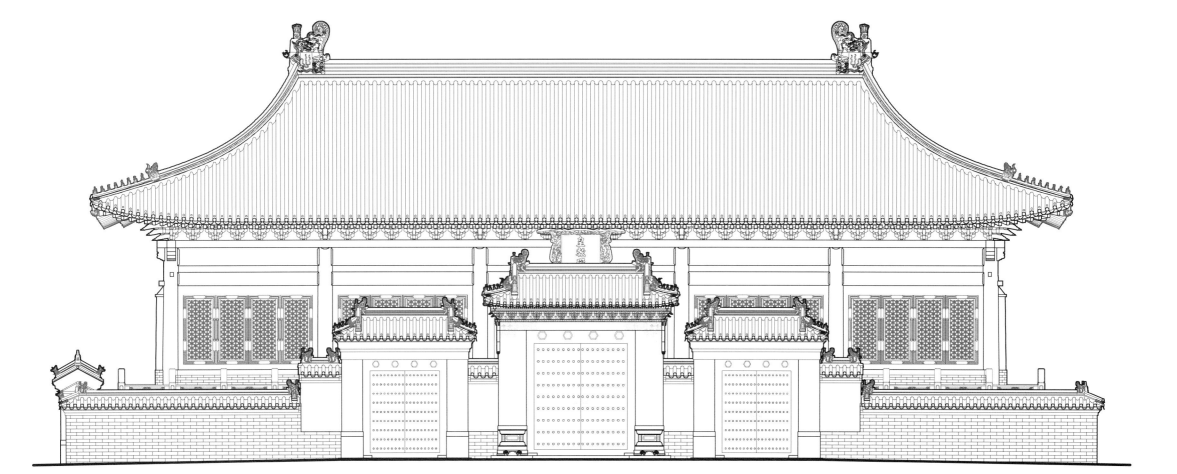

皇乾殿院落正立面图
Front elevation of courtyard of Hall of Imperial Zenith

0　1　　　　　5m

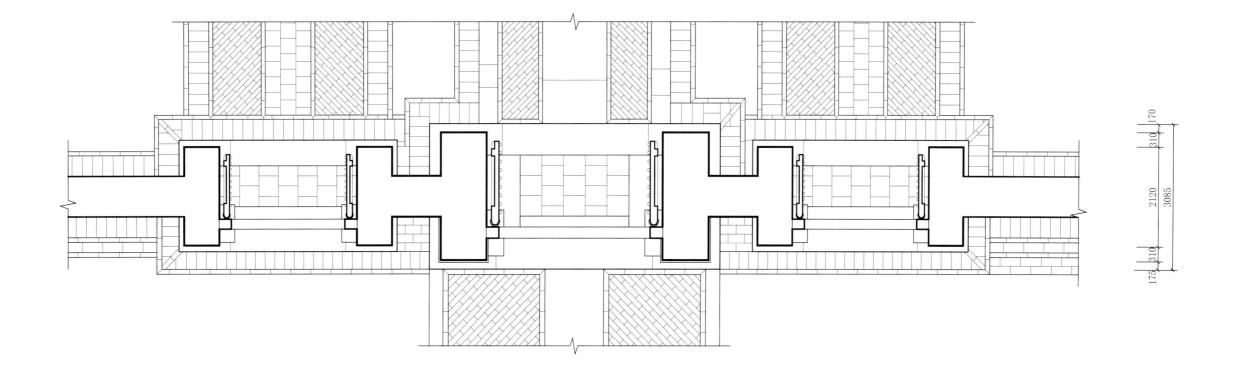

祈年殿北琉璃门平面图
Plan of north color-glazed gate of Hall of Prayer for Good Harvests

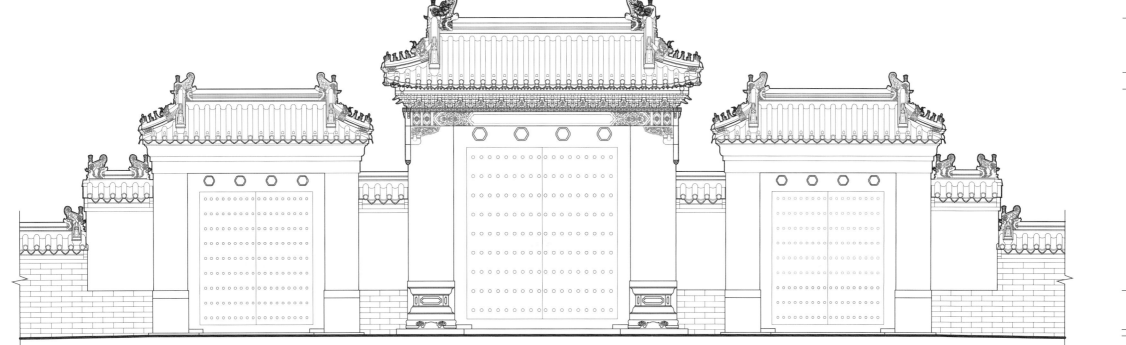

| 490 | 3750 | 1960 | 4870 | 1960 | 3750 | 490 |

17270

7.095
6.600
5.444
5.095
0.825
±0.000
-0.135

祈年殿北琉璃门正立面图
Front elevation of north color-glazed gate of Hall of Prayer for Good Harvests

0 1 2m

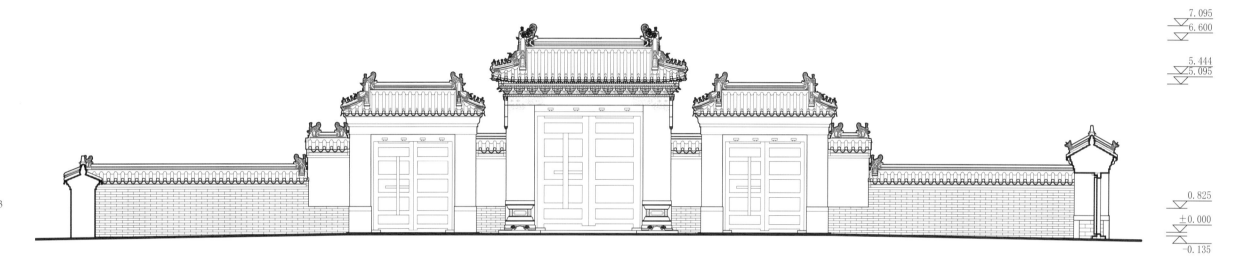

7.095
6.600

5.444
5.095

0.825
±0.000
-0.135

490 3750 1960 4870 1960 3750 490

17270

祈年殿北琉璃门背立面图及古稀门横剖面图

Rear elevation of north color-glazed gate and cross-section of Guxi door of Hall of Prayer for Good Harvests

0 1 3m

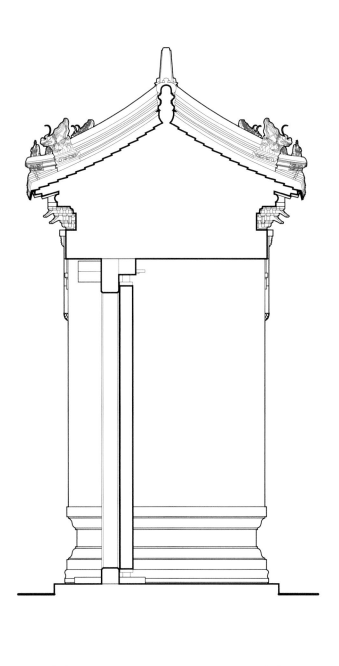

▽ 7.095

▽ 6.600

▽ 5.100

▽ 1.020

± 0.000

▽ −0.135

175	2740	170
	3085	

祈年殿北琉璃门中门侧立面图
Side elevation of central door of north color-glazed gate of Hall of Prayer for Good Harvests

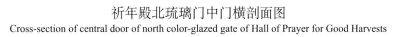

祈年殿北琉璃门中门横剖面图
Cross-section of central door of north color-glazed gate of Hall of Prayer for Good Harvests

0   1   2m

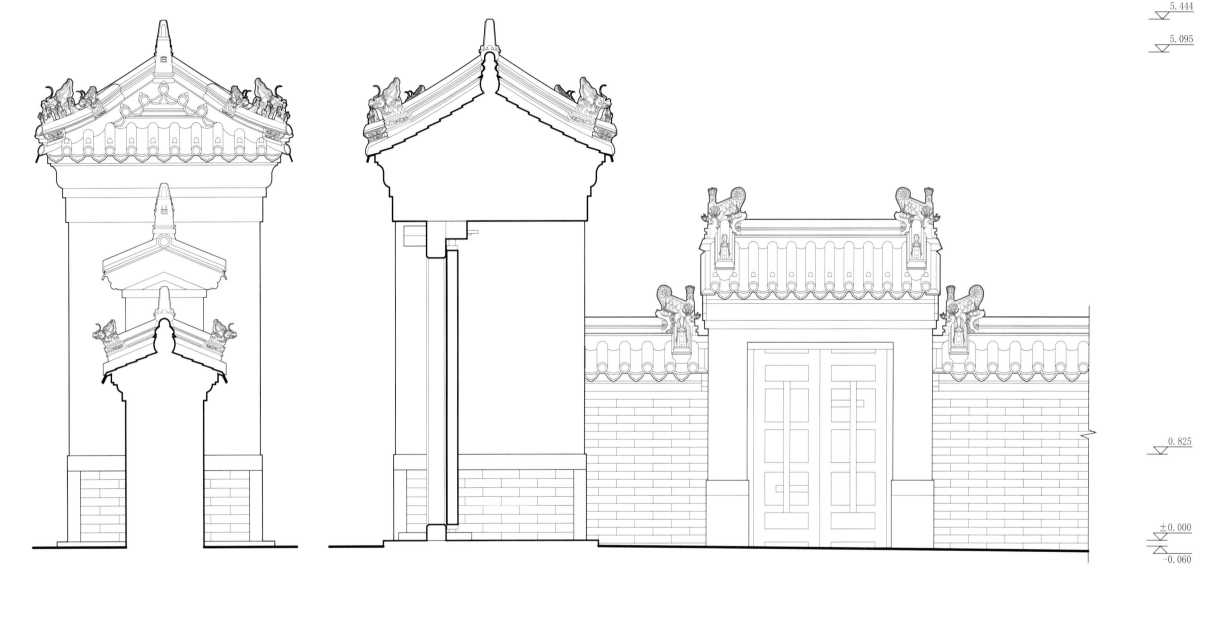

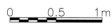

祈年殿北琉璃门边门侧立面图
Side elevation of side door of north color-glazed gate of Hall
of Prayer for Good Harvests

祈年殿北琉璃门边门横剖面图及古稀门背立面图
Cross-section of side door of north color-glazed gate and Rear elevation of Guxi door of
Hall of Prayer for Good Harvests

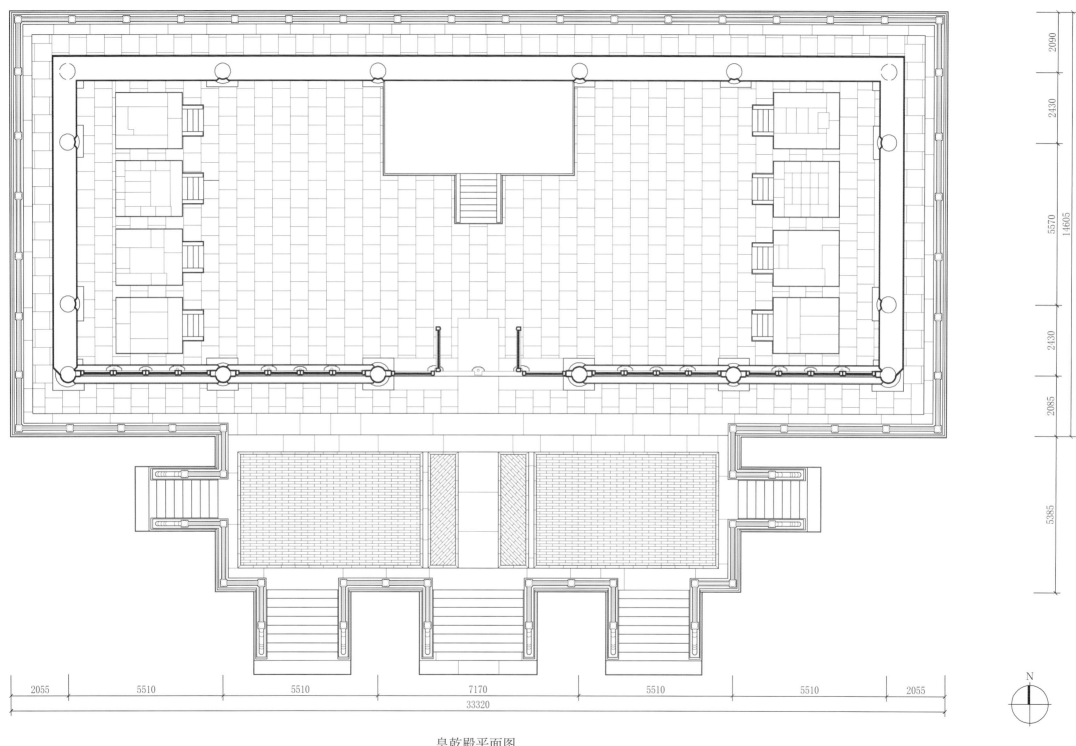

皇乾殿平面图
Plan of Hall of Imperial Zenith

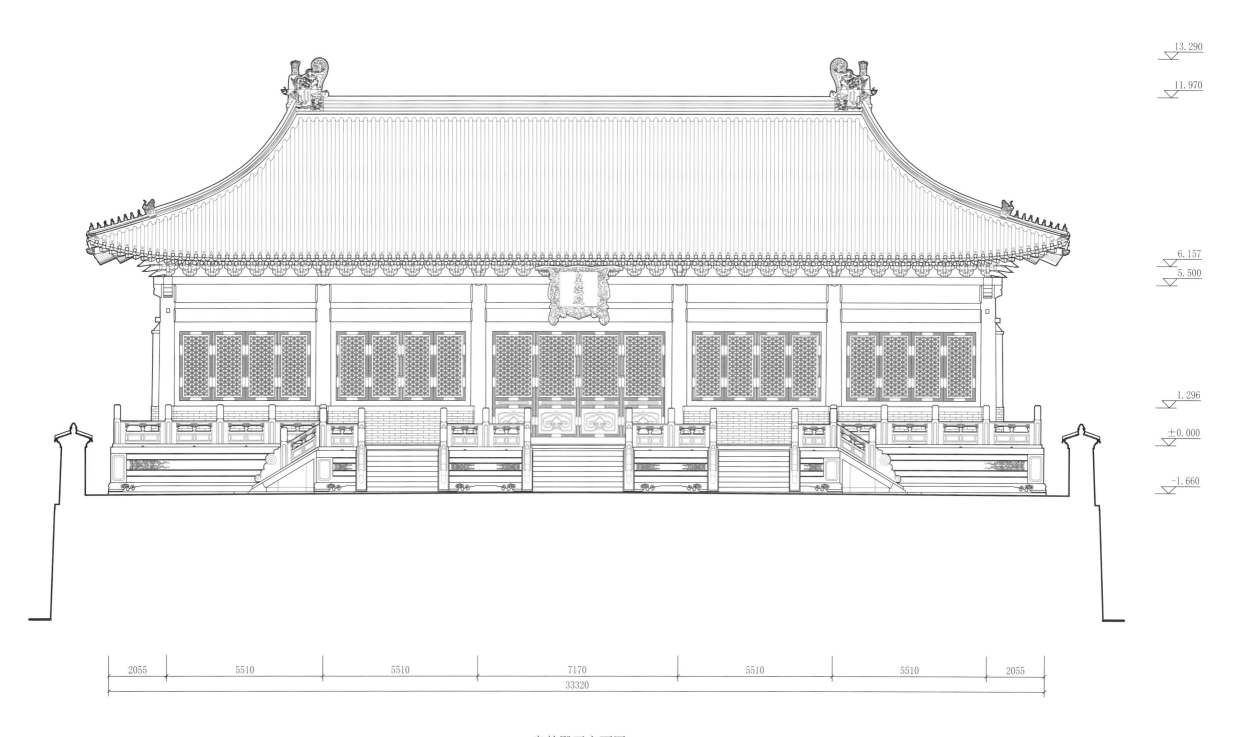

13.290

11.970

6.157

5.500

1.296

±0.000

-1.660

| 2055 | 5510 | 5510 | 7170 | 5510 | 5510 | 2055 |

33320

皇乾殿正立面图
Front elevation of Hall of Imperial Zenith

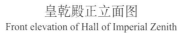

0 1 3m

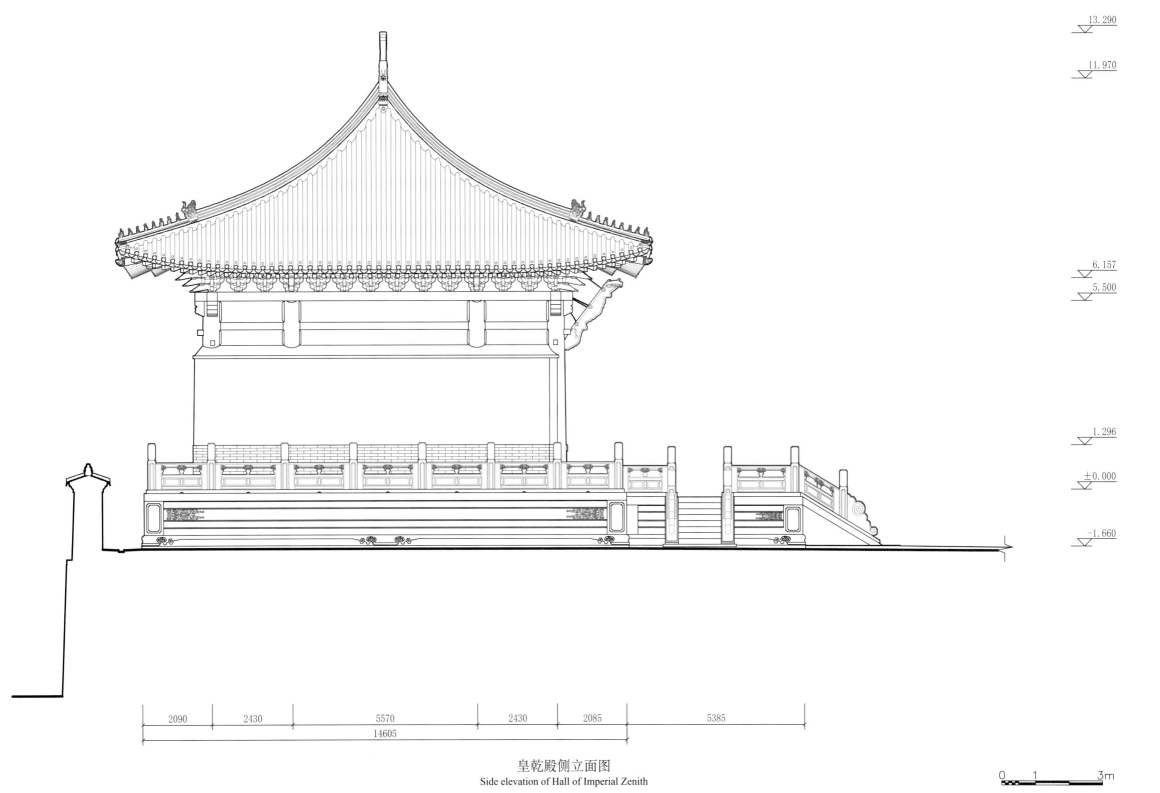

13.290

11.970

6.157

5.500

1.296

±0.000

-1.660

| 2090 | 2430 | 5570 | 2430 | 2085 | 5385 |

14605

皇乾殿侧立面图
Side elevation of Hall of Imperial Zenith

0 1 3m

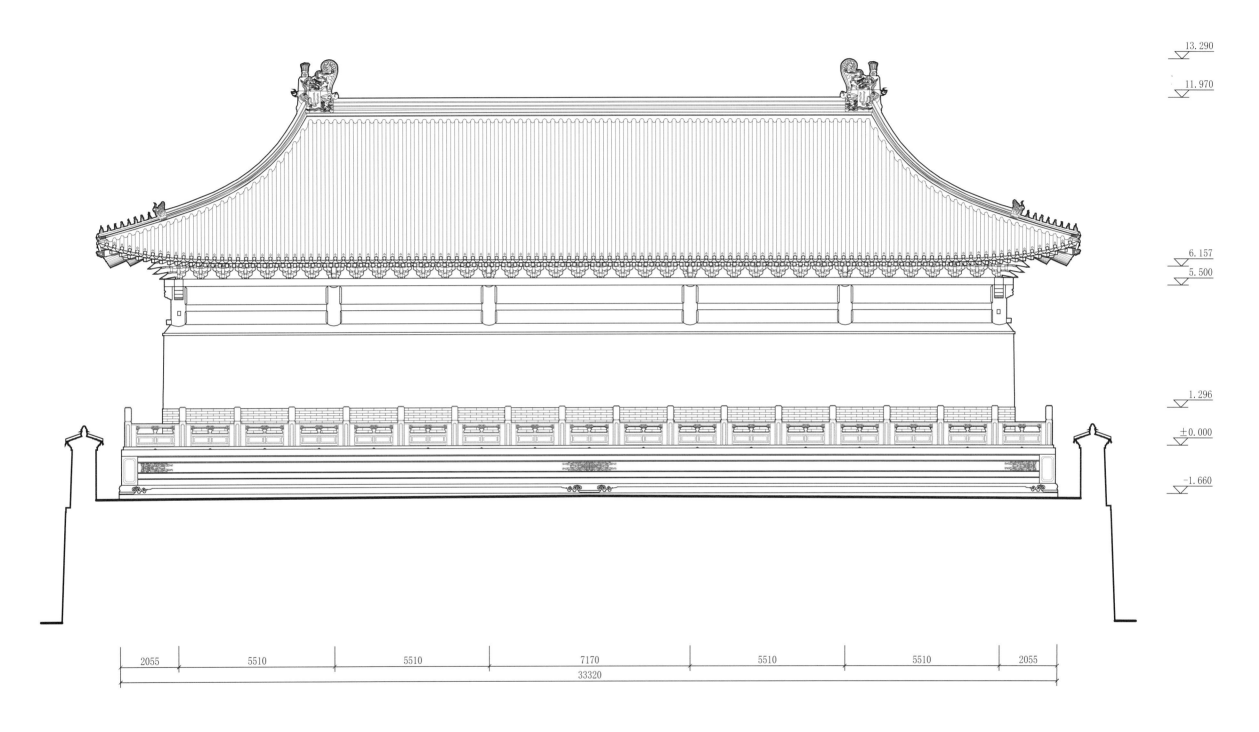

13. 290

11. 970

6. 157

5. 500

1. 296

±0. 000

-1. 660

2055	5510	5510	7170	5510	5510	2055

33320

皇乾殿背立面图
Rear elevation of Hall of Imperial Zenith

0 1 3m

13.290

11.970

6.157
5.500

1.296

±0.000

−1.660

| 2090 | 2430 | 5570 | 2430 | 2085 | 5385 |

14605

皇乾殿明间剖面图
Section of central-bay of Hall of Imperial Zenith

0 1 3m

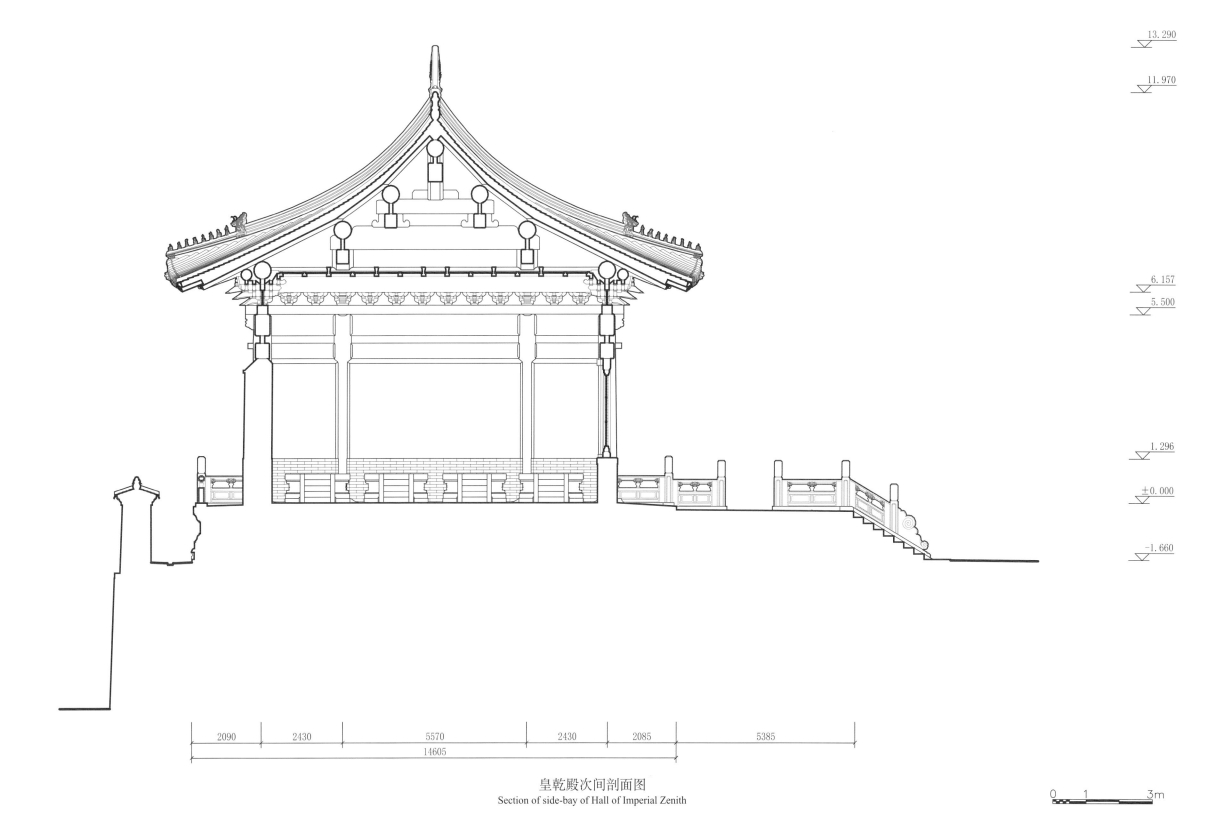

13.290

11.970

6.157

5.500

1.296

±0.000

-1.660

| 2090 | 2430 | 5570 | 2430 | 2085 | 5385 |

14605

皇乾殿次间剖面图
Section of side-bay of Hall of Imperial Zenith

0 1   3m

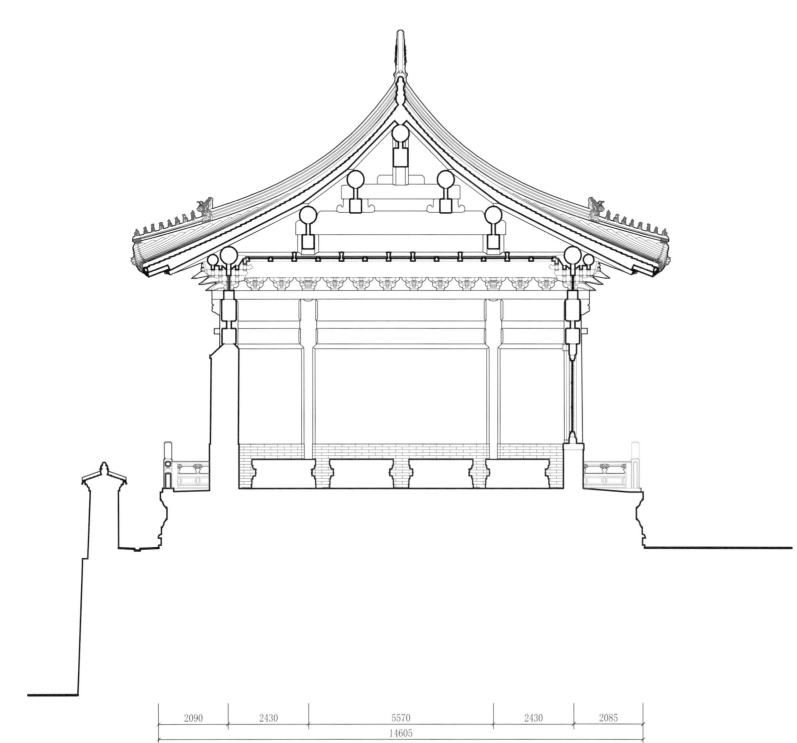

13.290

11.970

6.157

5.500

1.296

±0.000

-1.660

| 2090 | 2430 | 5570 | 2430 | 2085 |

14605

皇乾殿梢间剖面图
Section of second-to-last-bay of Hall of Imperial Zenith

0   1   3m

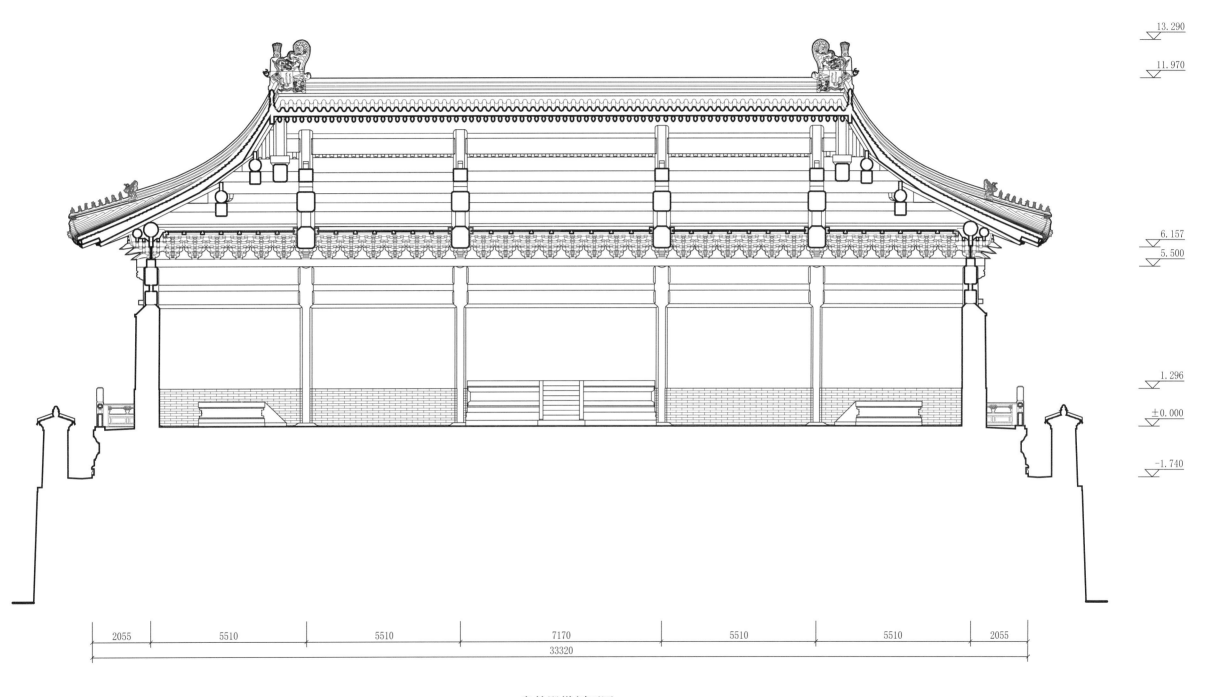

13.290

11.970

6.157
5.500

1.296

±0.000

-1.740

2055　5510　5510　7170　5510　5510　2055

33320

皇乾殿纵剖面图
Longitudinal section of Hall of Imperial Zenith

0　1　3m

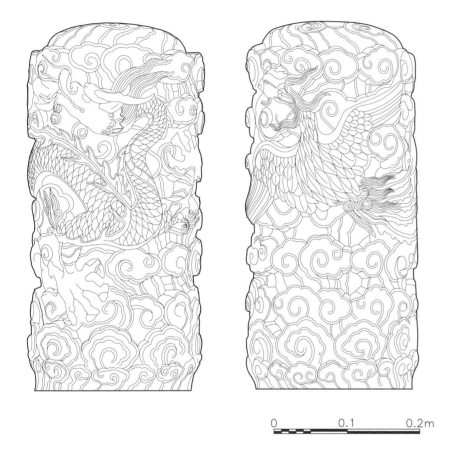

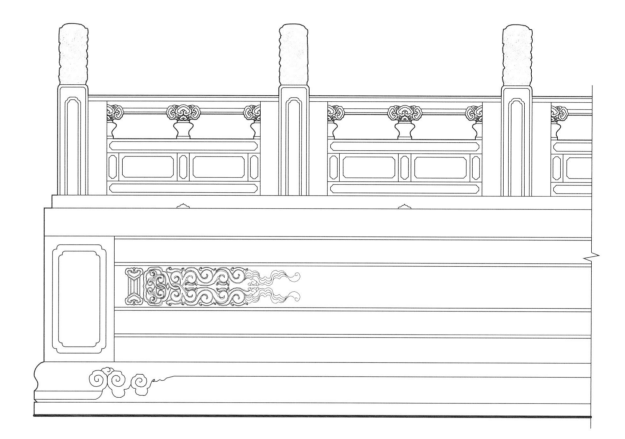

0   0.1   0.2m

皇乾殿望柱柱头大样图一
Wangzhu capital of Hall of Imperial Zenith

皇乾殿望柱柱头大样图二
Wangzhu capital of Hall of Imperial Zenith

皇乾殿台基立面大样图
Elevation of *Taiji* of Hall of Imperial Zenith

0   0.5   1m

1.560

±0.000

6786

皇乾殿神台正立面大样图
Front elevation of *Shentai* of Hall of Imperial Zenith

0 0.1　　0.5　　1m

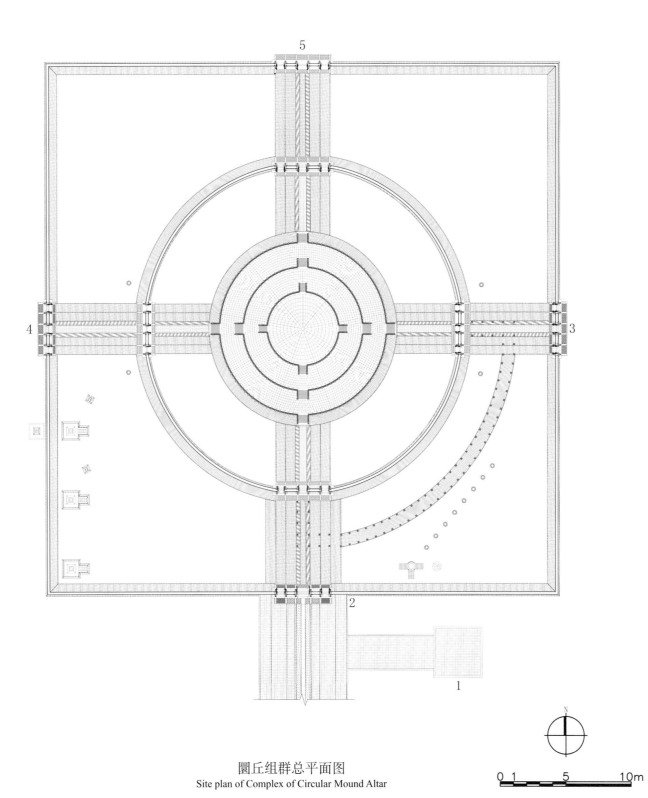

圜丘组群总平面图
Site plan of Complex of Circular Mound Altar

1 具服台　Dressing Terrace
2 外墙南棂星门　South Lingxing Gate from Outer Wall
3 外墙东棂星门　East Lingxing Gate from Outer Wall
4 外墙北棂星门　North Lingxing Gate from Outer Wall
5 外墙西棂星门　West Lingxing Gate from Outer Wall
6 燔柴炉　Firewood Stove
7 瘗坎　*Yikan*
8 望灯杆　Watching Light
9 内墙南棂星门　South Lingxing Gate from Inner Wall
10 内墙东棂星门　East Lingxing Gate from Inner Wall
11 内墙西棂星门　West Lingxing Gate from Inner Wall
12 内墙北棂星门　North Lingxing Gate from Inner Wall
13 圜丘　Circular Mound Altar

0 1　5　10m

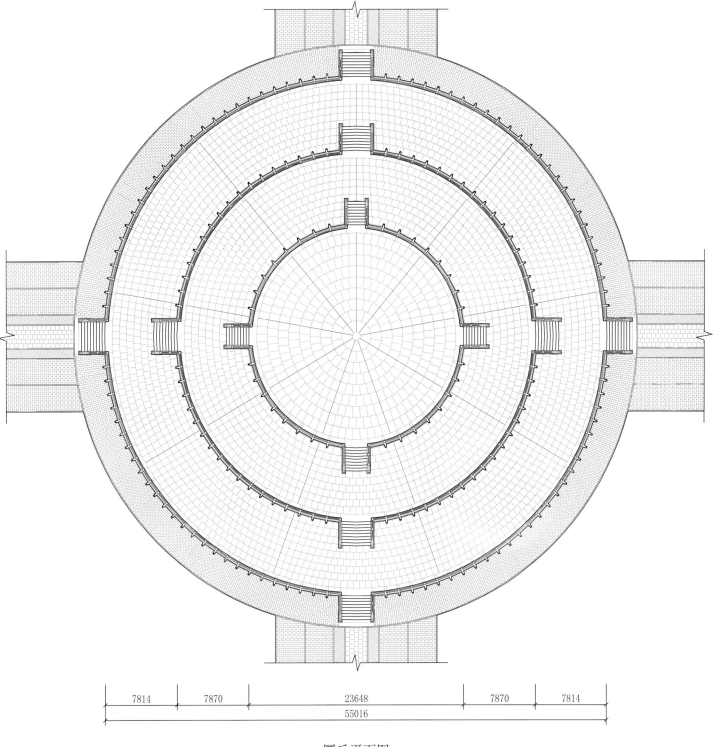

圜丘平面图
Site plan of Circular Mound Altar

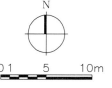

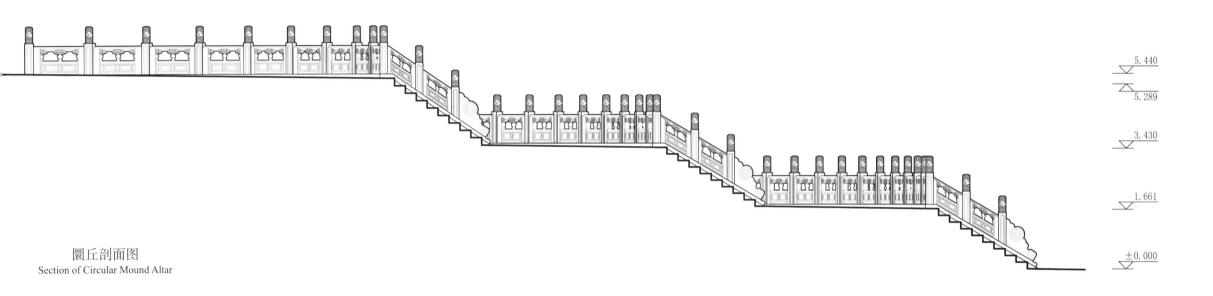

圜丘剖面图
Section of Circular Mound Altar

5.440
5.289
3.430
1.661
±0.000

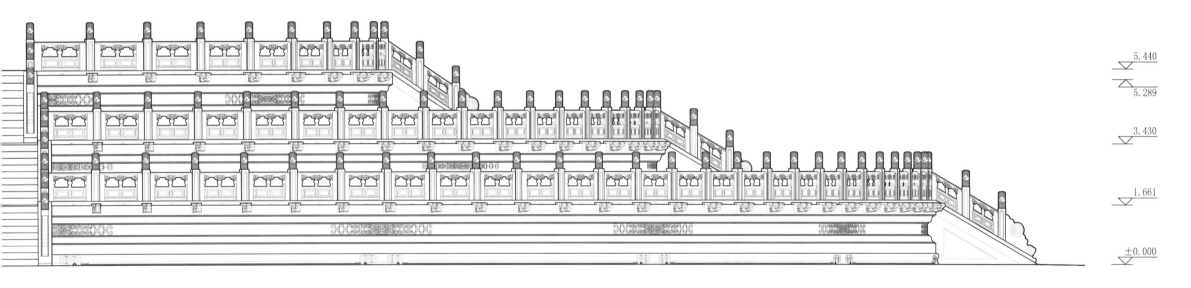

5.440
5.289
3.430
1.661
±0.000

10175　　　7849　　　7836

圜丘正立面图
Front elevation of Circular Mound Altar

0　1　3m

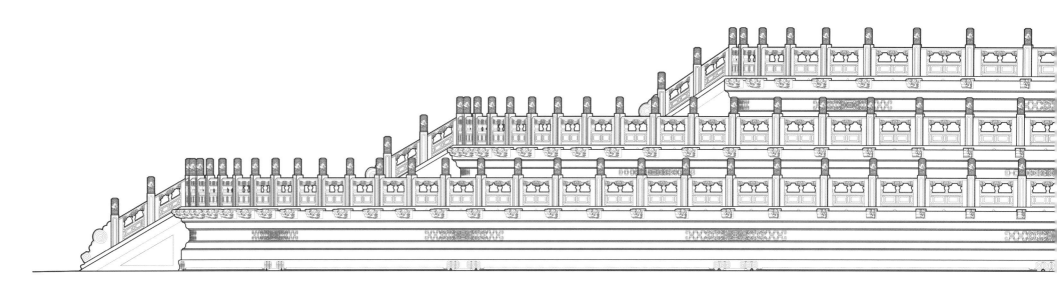

7836 7849 10175

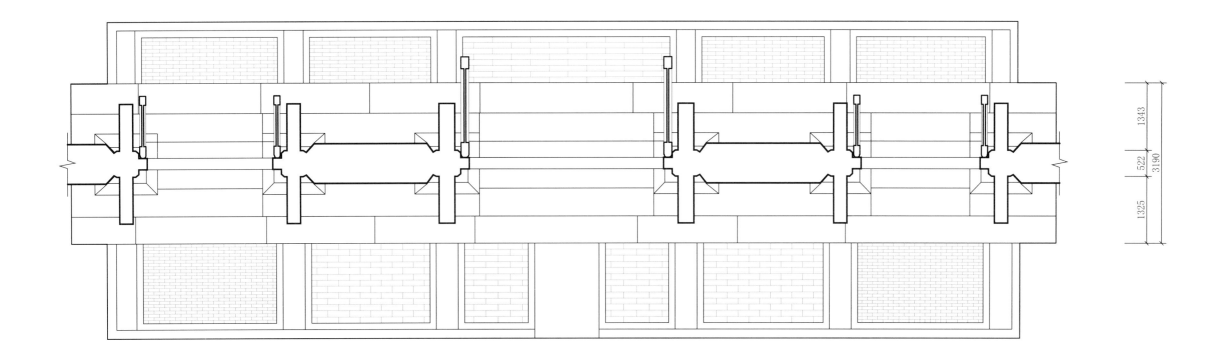

1343

522

3190

1325

| 1124 | 3442 | 3143 | 4908 | 3173 | 3288 | 1124 |

20202

圜丘外墙南棂星门平面图
Plan of South Lingxing Gate from outer wall of Circular Mound Altar

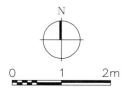

7.394

5.460

4.294

2.972

2.057

±0.000

-0.110

1124　　　3442　　　3143　　　4908　　　3173　　　3288　　1124

20202

圜丘外壝南棂星门正立面图

Front elevation of South Lingxing Gate from outer wall of Circular Mound Altar

0　　1　　2m

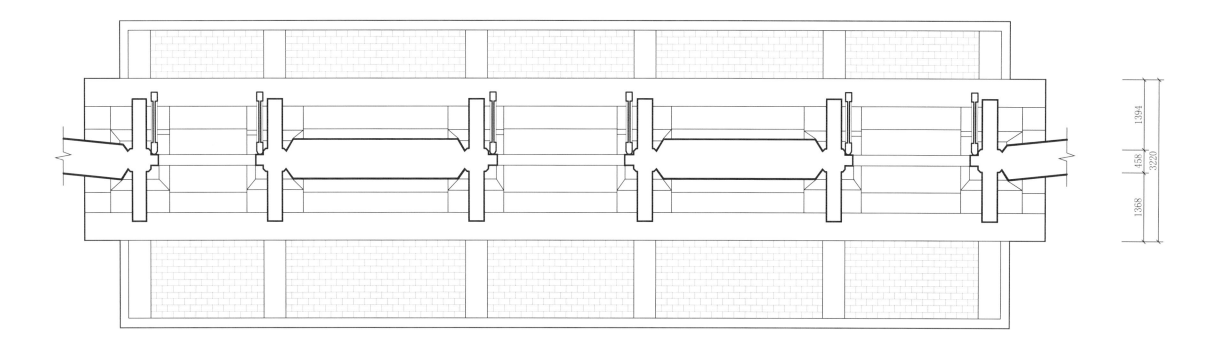

圜丘内壝南棂星门平面图
Plan of South Lingxing Gate from inner wall of Circular Mound Altar

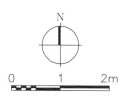

0　　1　　2m

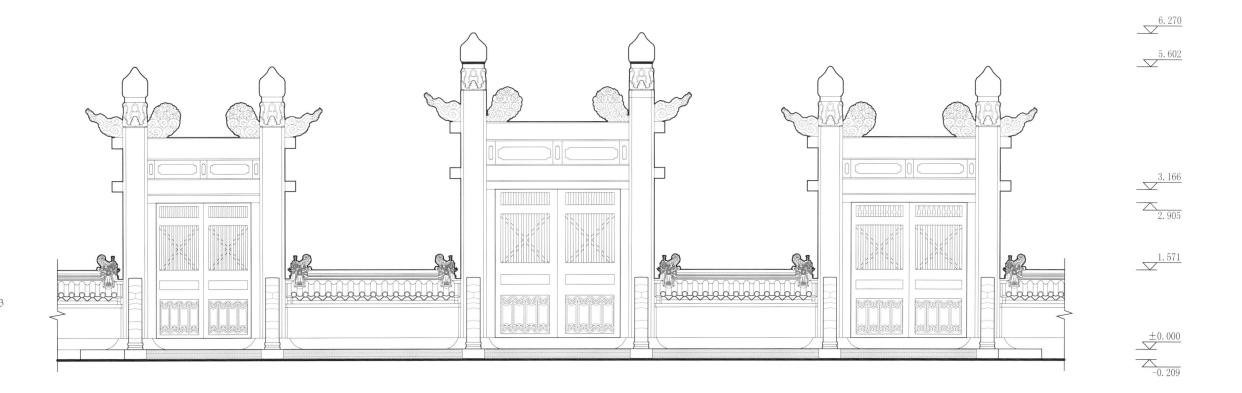

6.270

5.602

3.166

2.905

1.571

±0.000

-0.209

1125　2779　4141　3454　3886　3191　1125

19701

-0.209

圜丘内壝南棂星门正立图
Front elevation of South Lingxing Gate from inner wall of Circular Mound Altar

0　1　2m

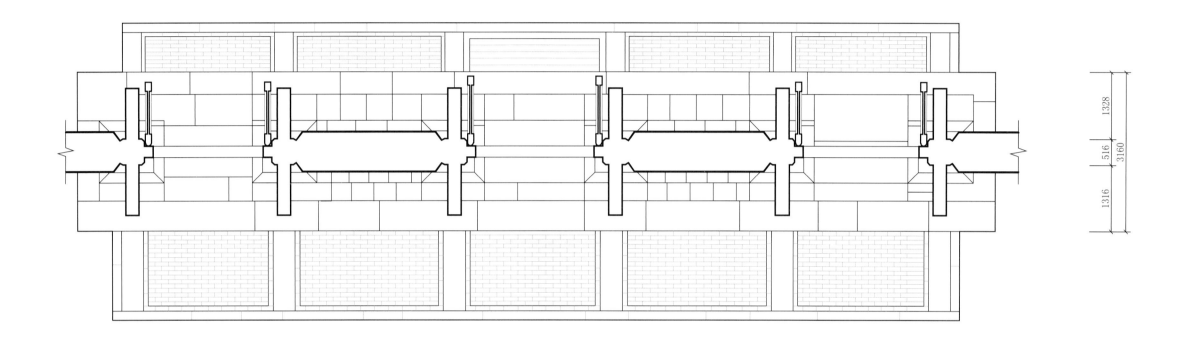

圜丘外墙西棂星门平面图
Plan of West Lingxing Gate from outer wall of Circular Mound Altar

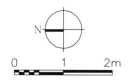

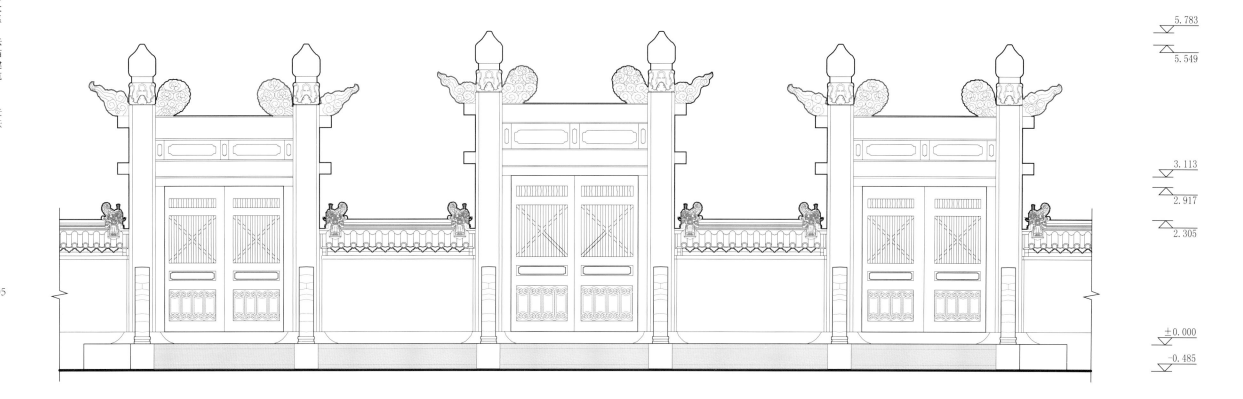

5.783
5.549
3.113
2.917
2.305
±0.000
-0.485

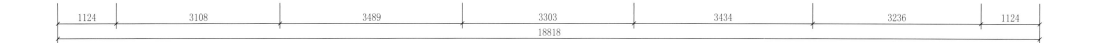

1124　　3108　　3489　　3303　　3434　　3236　　1124
18818

圜丘外墙西棂星门正立面图
Front elevation of West Lingxing Gate from outer wall of Circular Mound Altar

0　　1　　2m

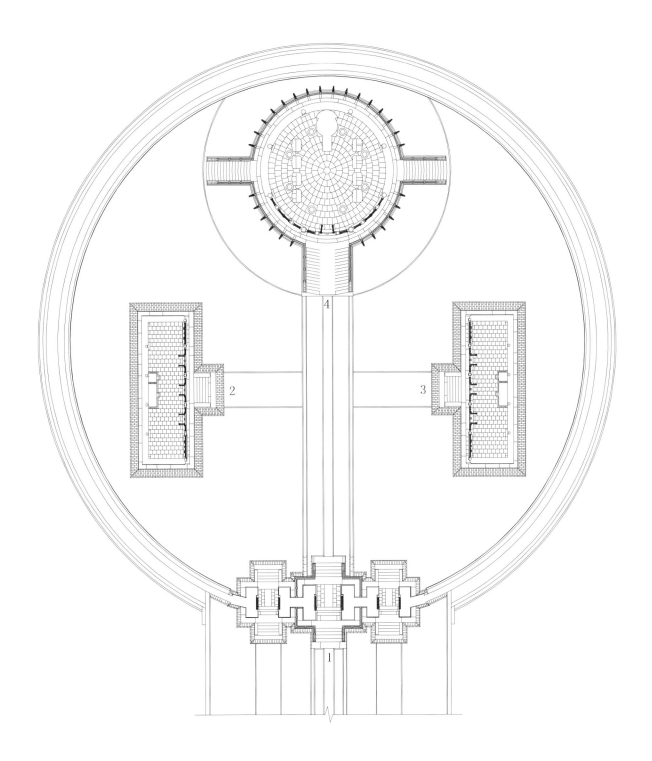

1　琉璃花门　Color-glazed Huamen
2　西配殿　West Side Hall
3　东配殿　East Side Hall
4　皇穹宇　Imperial Vault of Heaven

皇穹宇组群总平面图
Site plan of Complex of Imperial Vault of Heaven

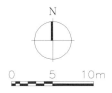

N

0　　5　　10m

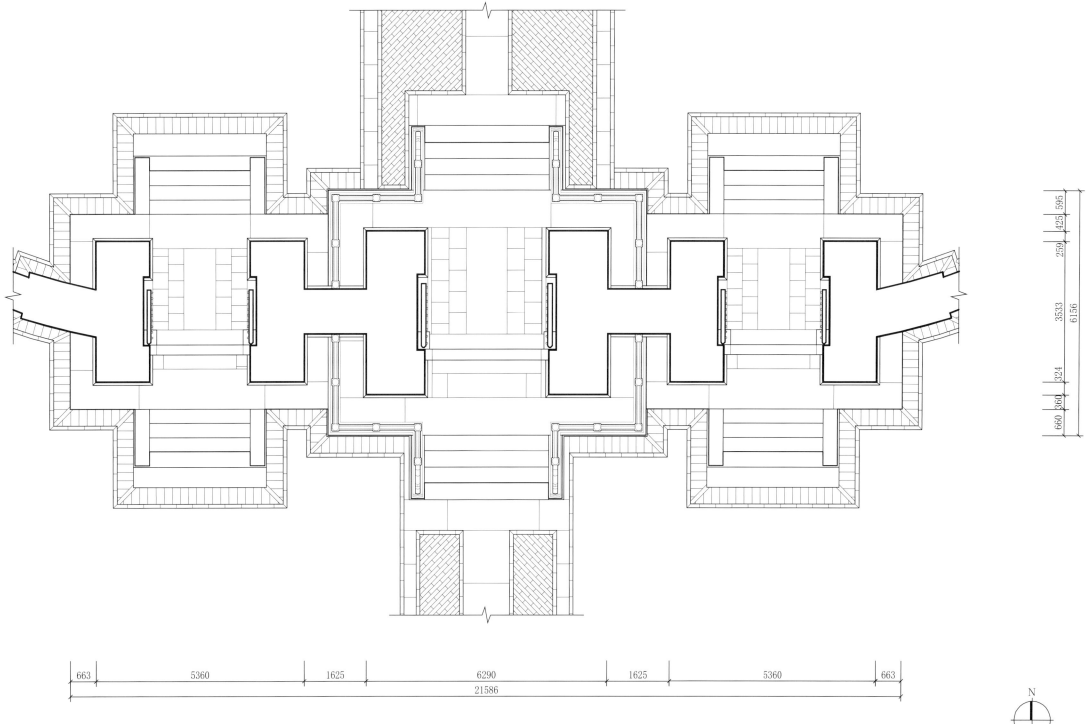

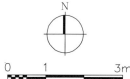

皇穹宇琉璃花门平面图
Plan of color-glazed *huamen* of Imperial Vault of Heaven

663 5360 1625 6290 1625 5360 663
21586

595 425 259 3533 6156 324 360 660

0 1 3m

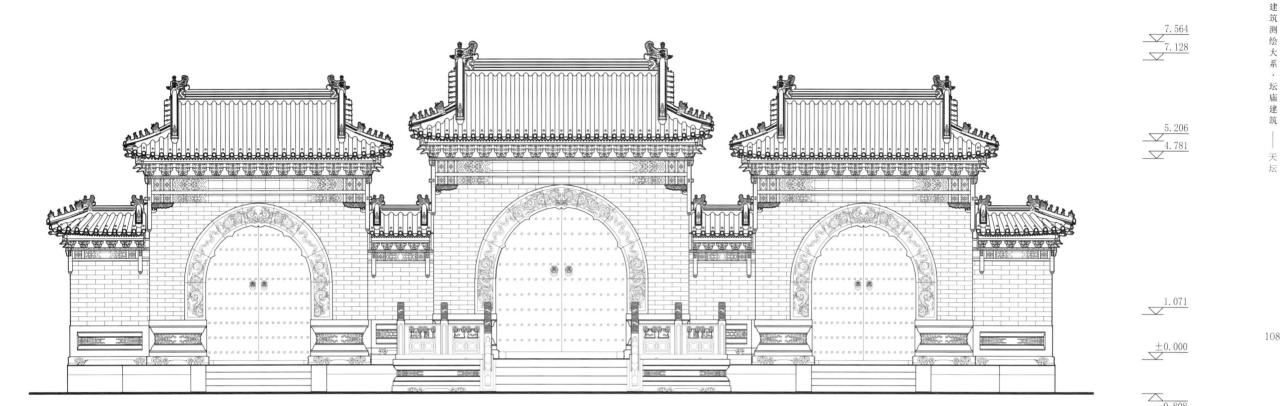

7.564
7.128

5.206
4.781

1.071

±0.000

−0.808

| 663 | 5360 | 1625 | 6290 | 1625 | 5360 | 663 |

21586

皇穹宇琉璃花门正立面图
Front elevation of color-glazed *huamen* of Imperial Vault of Heaven

0 1 3m

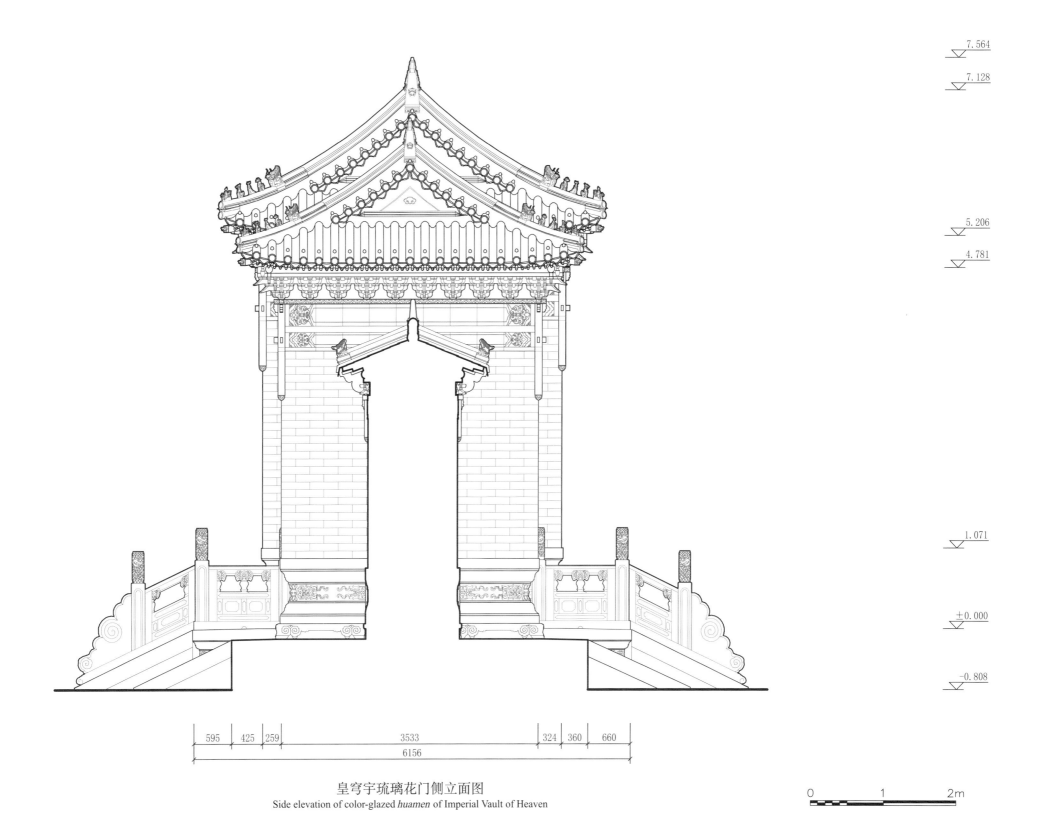

7.564

7.128

5.206

4.781

1.071

±0.000

-0.808

| 595 | 425 | 259 | 3533 | 324 | 360 | 660 |

6156

皇穹宇琉璃花门侧立面图
Side elevation of color-glazed *huamen* of Imperial Vault of Heaven

0   1   2m

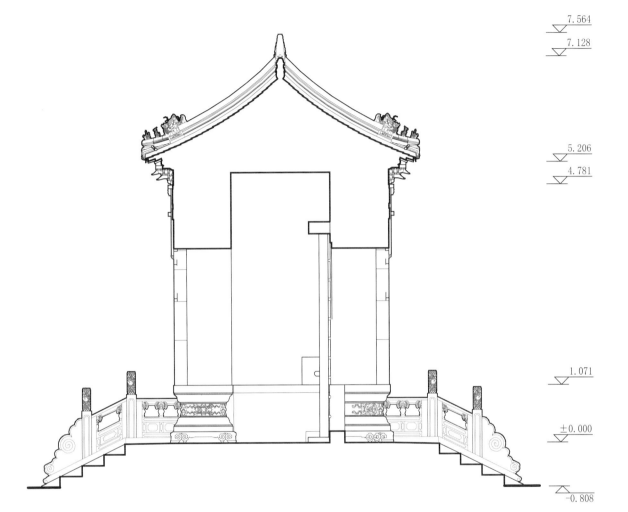

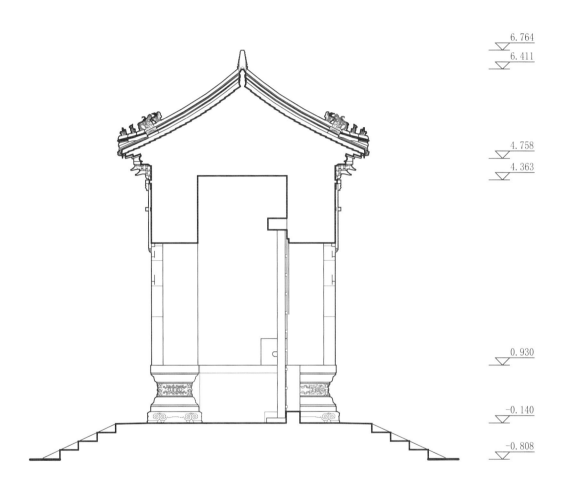

7.564

7.128

6.764

6.411

5.206

4.781

4.758

4.363

1.071

0.930

±0.000

-0.140

-0.808

-0.808

955	4116	955
	6026	

684	3533	684
	4901	

皇穹宇琉璃花门中门剖面图
Section of central door of color-glazed *huamen* of Imperial Vault of Heaven

皇穹宇琉璃花门边门剖面图
Section of side door of color-glazed *huamen* of Imperial Vault of Heaven

0　　　1　　　2m

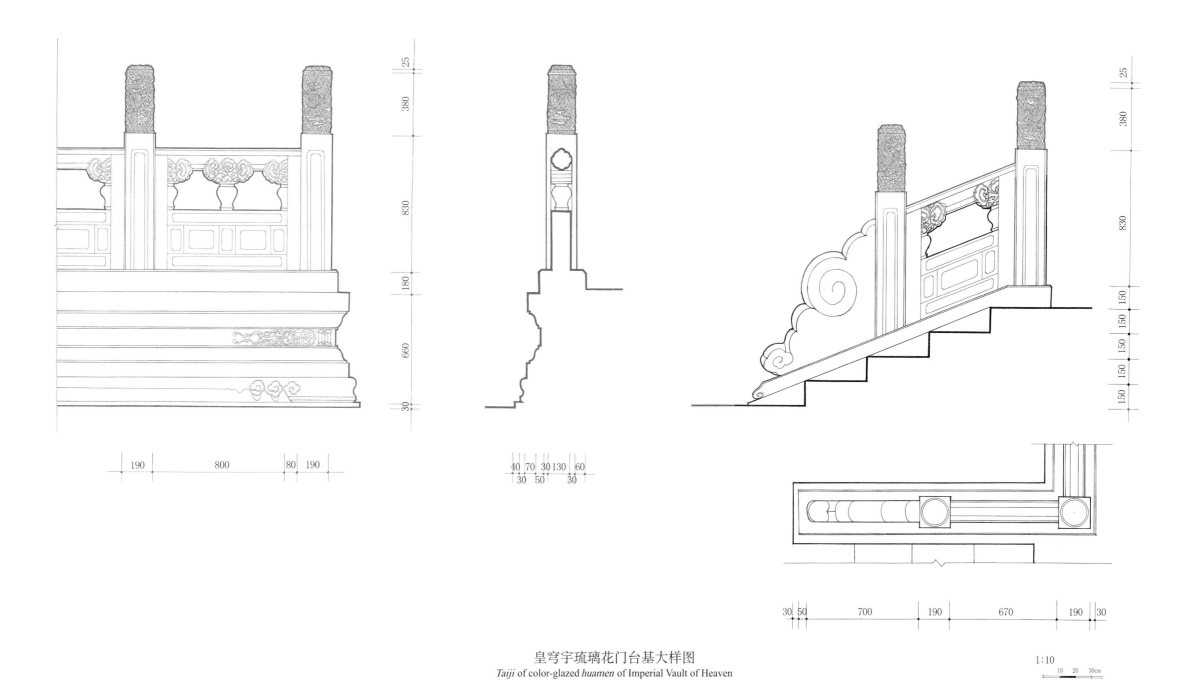

皇穹宇琉璃花门台基大样图
Taiji of color-glazed *huamen* of Imperial Vault of Heaven

1:10

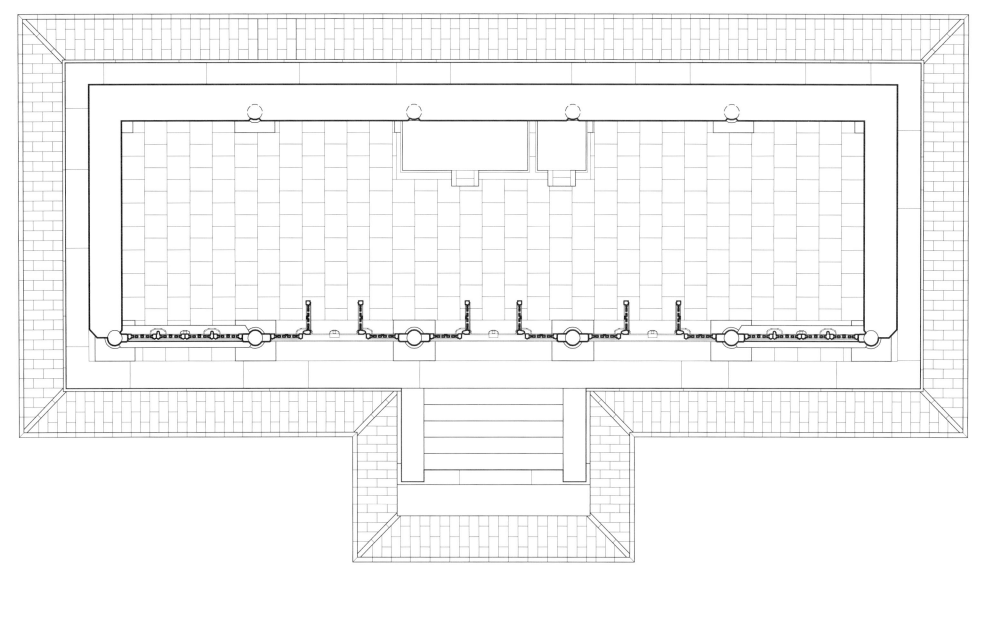

皇穹宇西配殿平面图
Plan of west side hall of Imperial Vault of Heaven

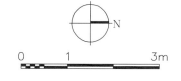

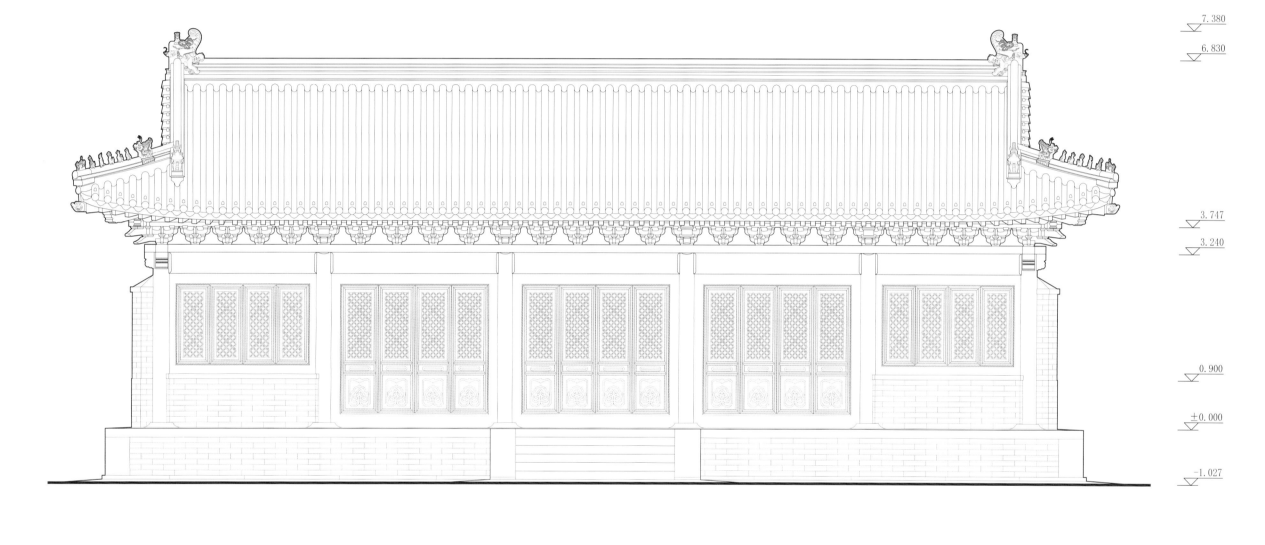

| 1075 | 3070 | 3480 | 3480 | 3480 | 3070 | 1075 |

18730

皇穹宇西配殿正立面图
Front elevation of west side hall of Imperial Vault of Heaven

0　　1　　2m

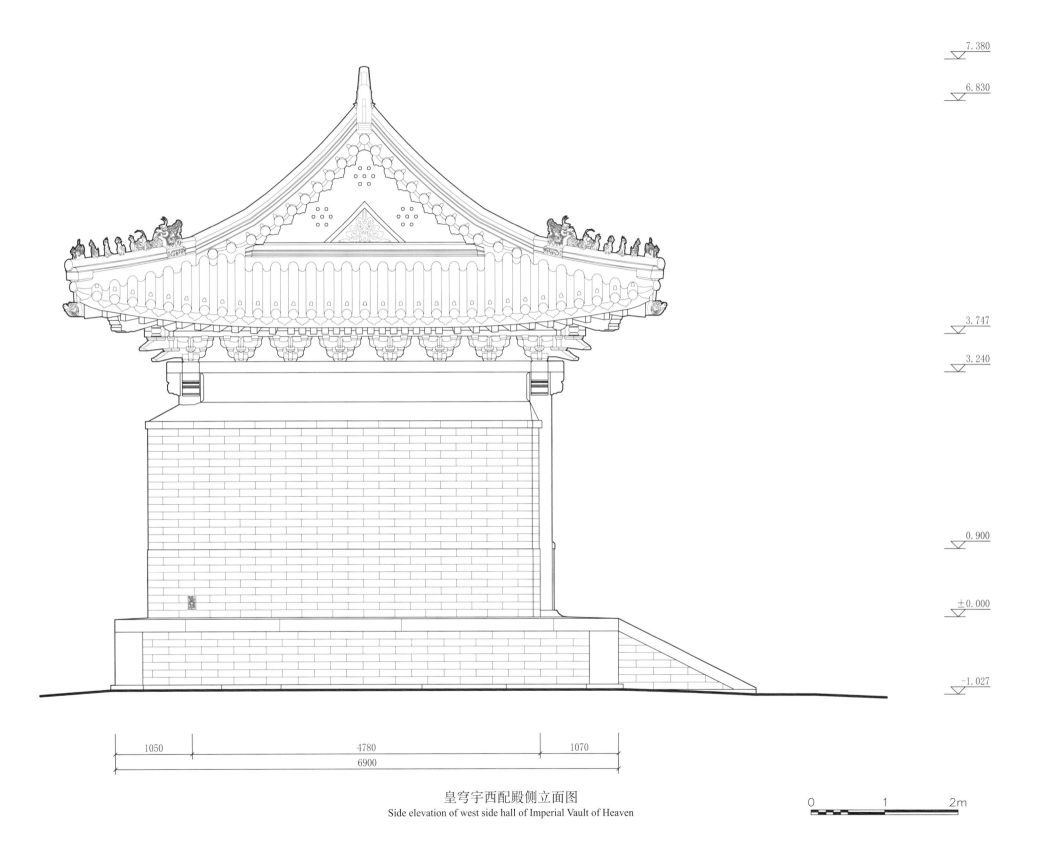

7.380

6.830

3.747

3.240

0.900

±0.000

-1.027

1050　　4780　　1070

6900

皇穹宇西配殿侧立面图
Side elevation of west side hall of Imperial Vault of Heaven

0　　1　　2m

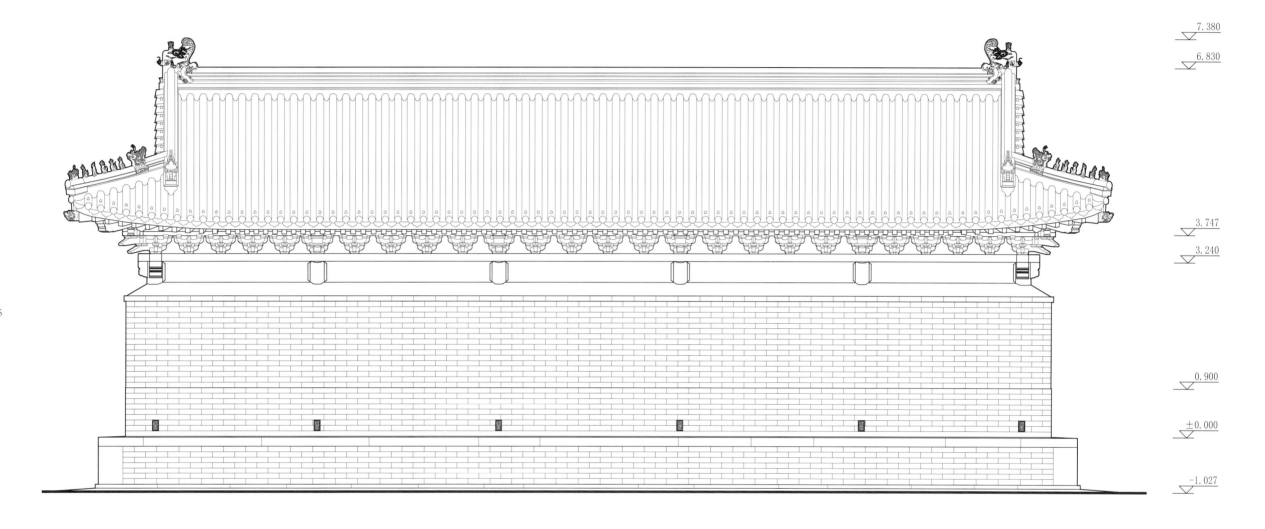

7.380

6.830

3.747

3.240

0.900

±0.000

-1.027

| 1075 | 3070 | 3480 | 3480 | 3480 | 3070 | 1075 |

18730

皇穹宇西配殿背立面图
Rear elevation of west side hall of Imperial Vault of Heaven

0　　1　　2m

7.380

6.830

3.747

3.240

0.900

±0.000

-1.027

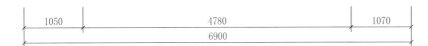

| 1050 | 4780 | 1070 |
6900

| 1050 | 4780 | 1070 |
6900

皇穹宇西配殿明间剖面图
Section of central-bay of west side hall of Imperial Vault of Heaven

皇穹宇西配殿梢间剖面图
Section of second-to-last-bay of west side hall of Imperial Vault of Heaven

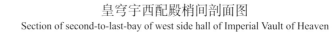 

0   1   2m

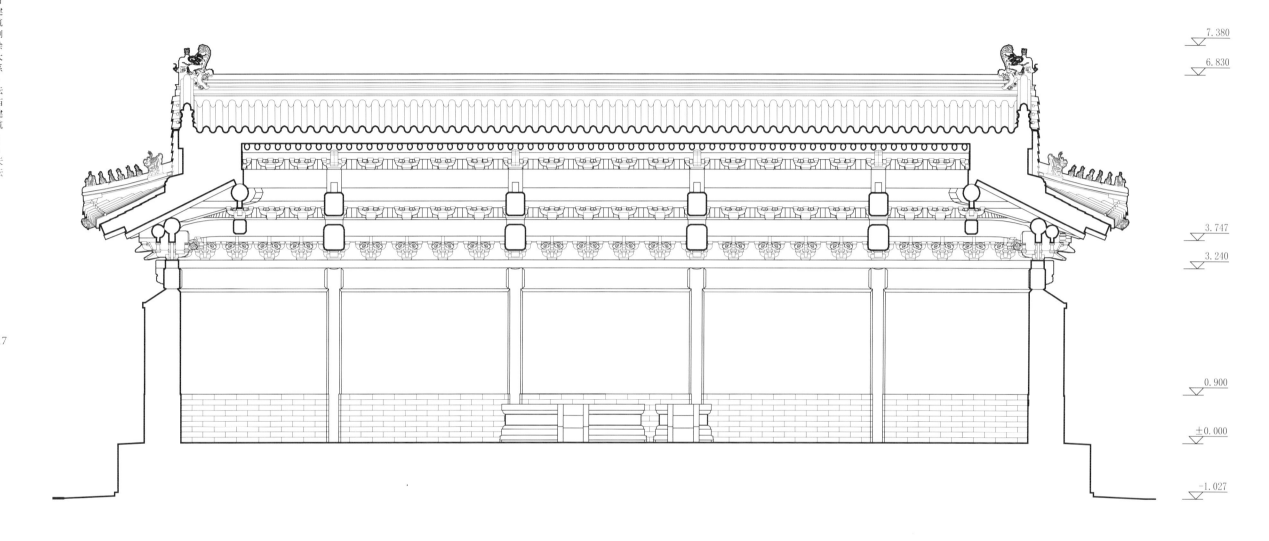

7.380

6.830

3.747

3.240

0.900

±0.000

-1.027

| 1075 | 3070 | 3480 | 3480 | 3480 | 3070 | 1075 |

18730

皇穹宇西配殿纵剖面图
Longitudinal section of west side hall of Imperial Vault of Heaven

0　　1　　2m

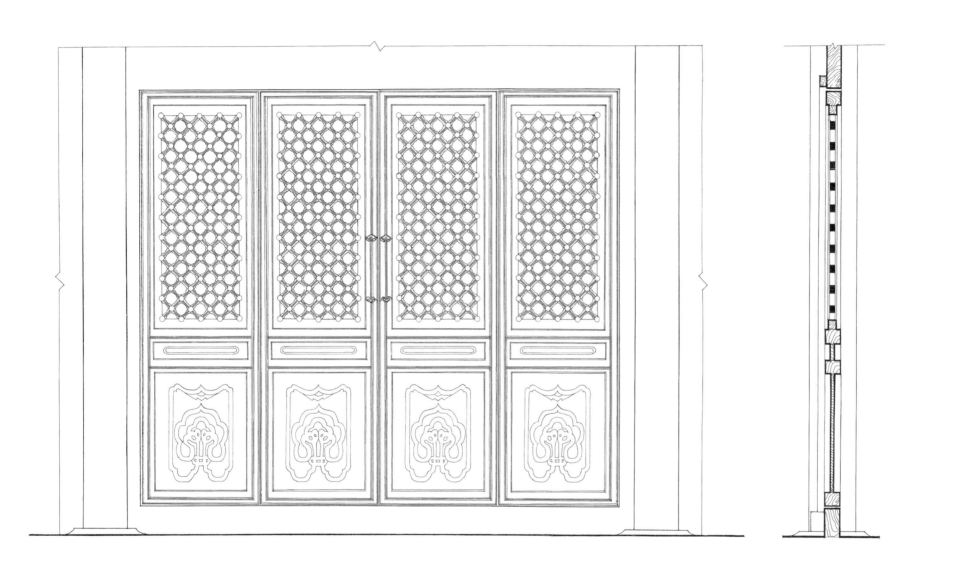

皇穹宇西配殿隔扇大样图
Geshan of west side hall of Imperial Vault of Heaven

0　　　0.5　　　1m

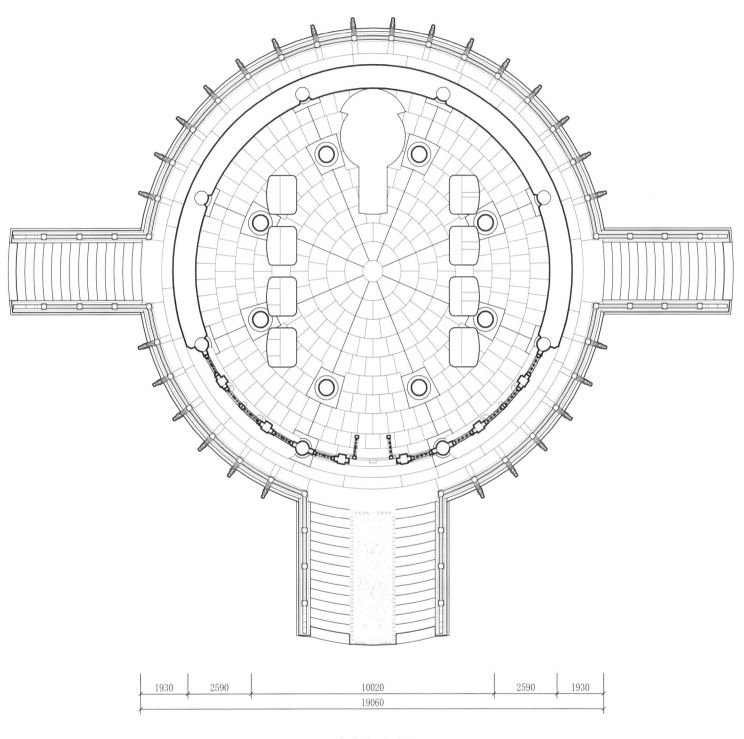

皇穹宇平面图
Plan of Imperial Vault of Heaven

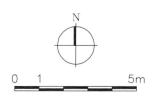

0 1   5m

16.715

14.354

6.240
5.705

1.460

±0.000

-2.900

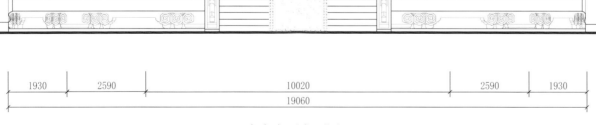

| 1930 | 2590 | 10020 | 2590 | 1930 |

19060

皇穹宇正立面图
Front elevation of Imperial Vault of Heaven

0   1   3m

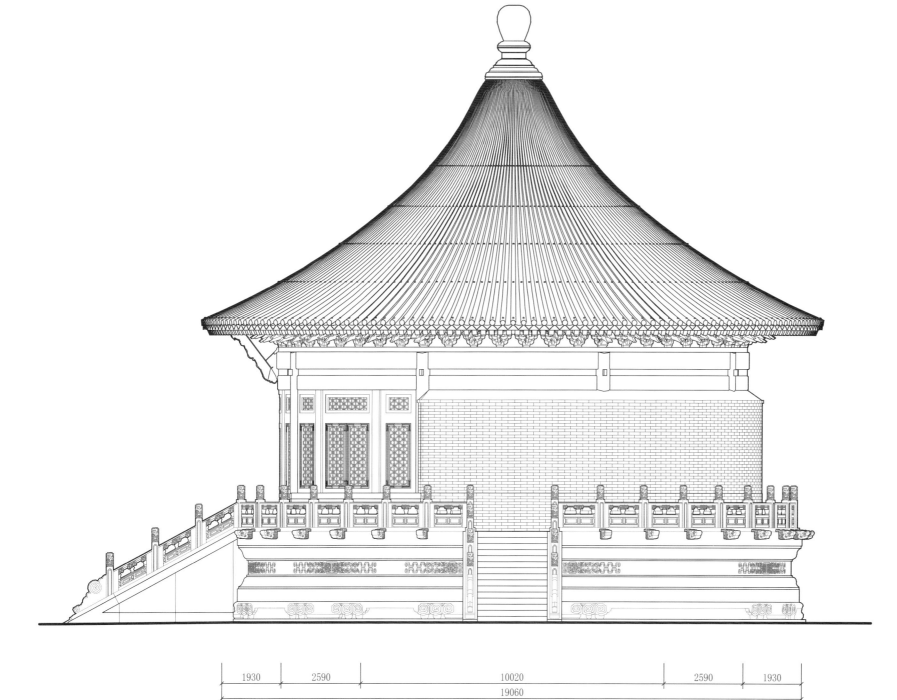

16.715

14.354

6.240

5.705

1.460

±0.000

-2.900

| 1930 | 2590 | 10020 | 2590 | 1930 |

19060

皇穹宇侧立面图
Side elevation of Imperial Vault of Heaven

0　1　　　3m

16.715

14.354

6.240
5.705

1.460

±0.000

-2.900

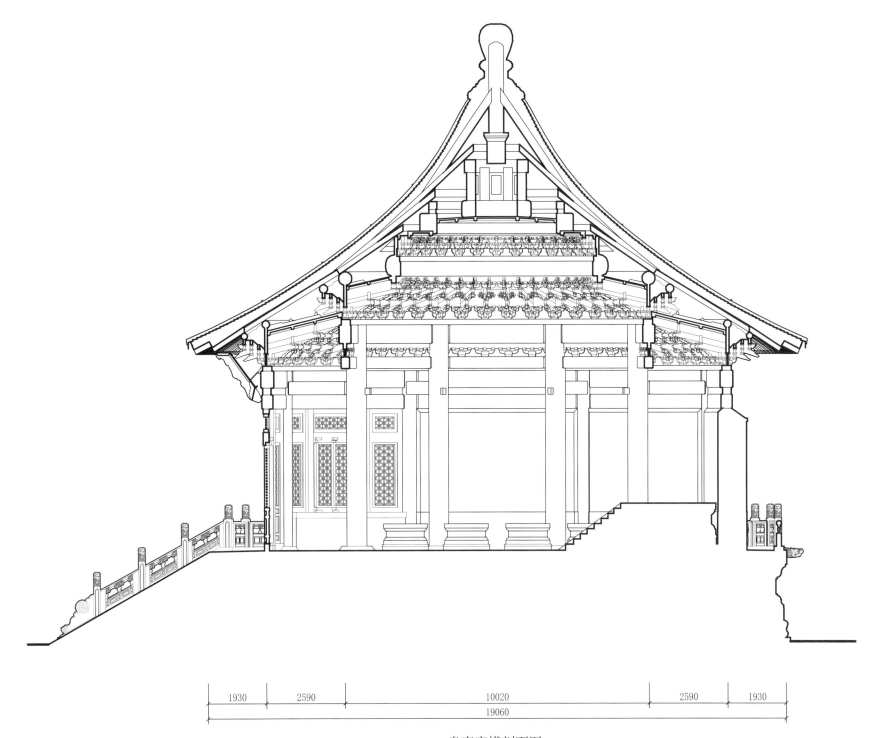

| 1930 | 2590 | 10020 | 2590 | 1930 |

19060

皇穹宇横剖面图
Cross-section of Imperial Vault of Heaven

0 1 3m

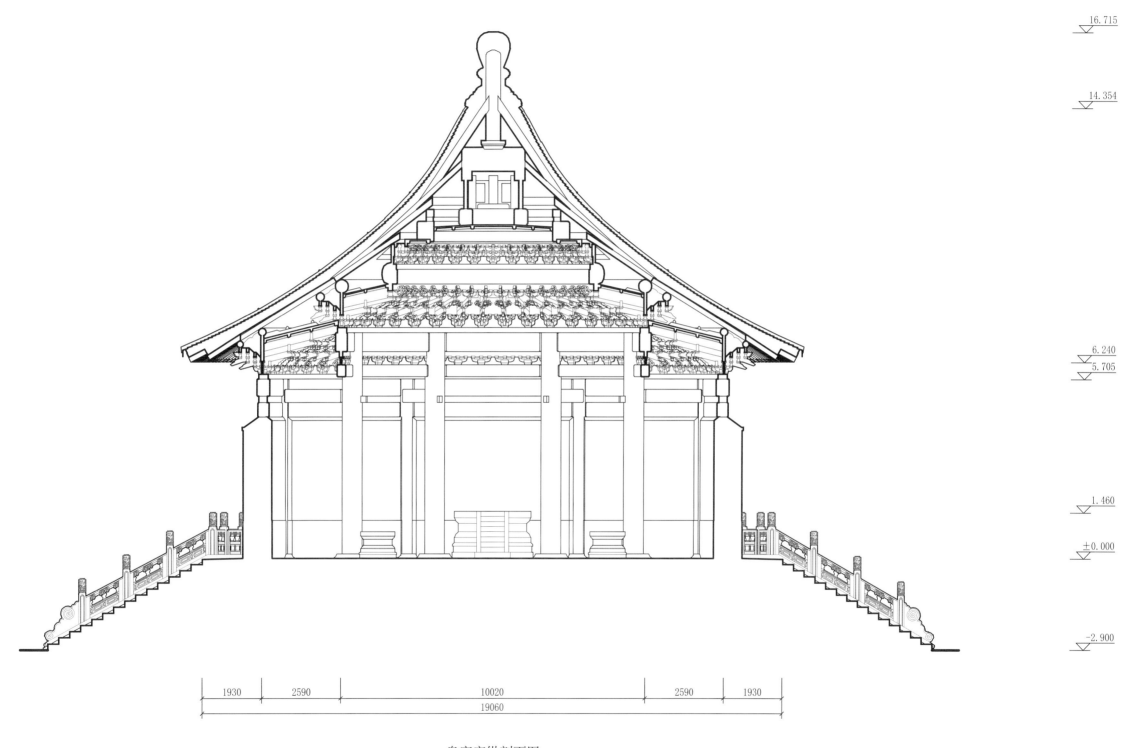

16.715

14.354

6.240
5.705

1.460

±0.000

-2.900

| 1930 | 2590 | 10020 | 2590 | 1930 |

19060

皇穹宇纵剖面图
Longitudinal section of Imperial Vault of Heaven

0　　1　　　　3m

皇穹宇丹陛大样图
Danbi of Imperial Vault of Heaven

0　　0.2　　0.4m

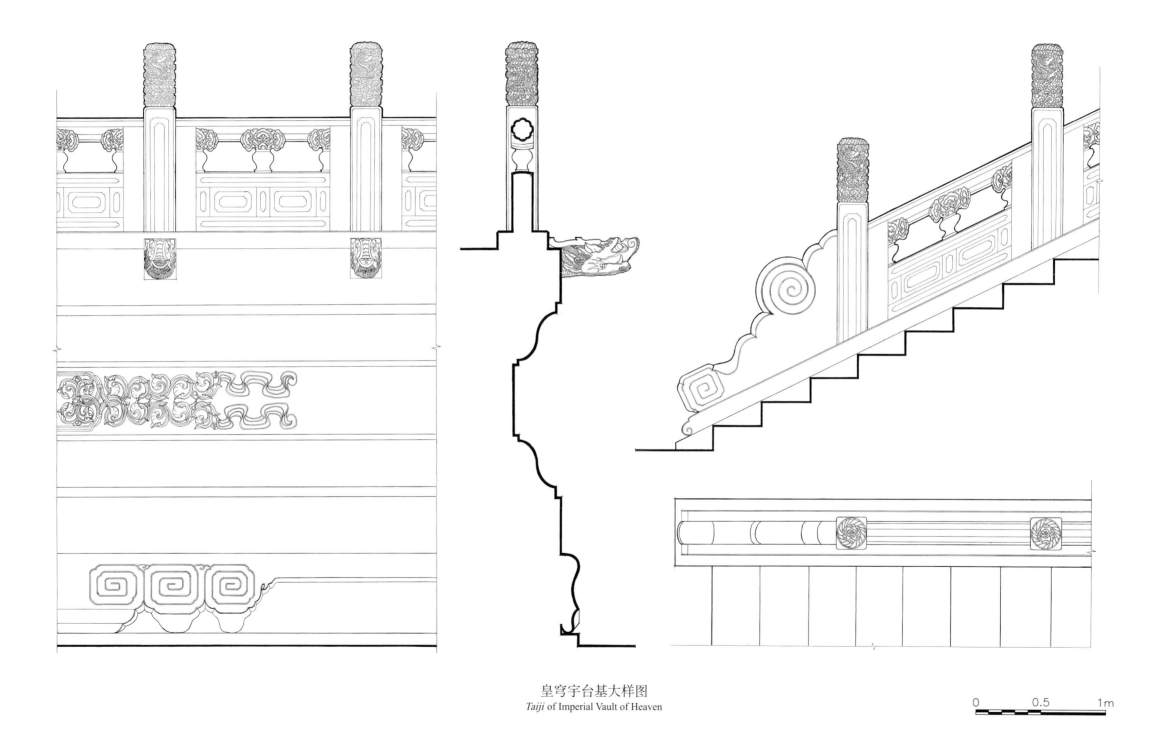

皇穹宇台基大样图
Taiji of Imperial Vault of Heaven

0　　　0.5　　　1m

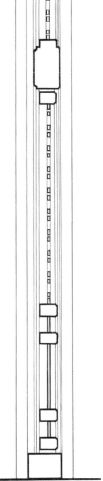

皇穹宇隔扇大样图
Geshan of Imperial Vault of Heaven

0   0.5   1m

斋宫组群
Complex of Palace
of Abstinence

1　东外门　East Outer Gate
2　钟楼　Bell Tower
3　东内门　East Inner Gate
4　北外门　North Outer Gate
5　北内门　North Inner Gate
6　南外门　South Outer Gate
7　南内门　South Inner Gate
8　正殿　Main Hall
9　值守房　*Zhishou* House
10　典守房　*Dianshou* House
11　垂花门　*Chuihuamen*
12　寝殿　Sleeping Hall
13　左随事房　Left *Suishi* House
14　右随事房　Right *Suishi* House

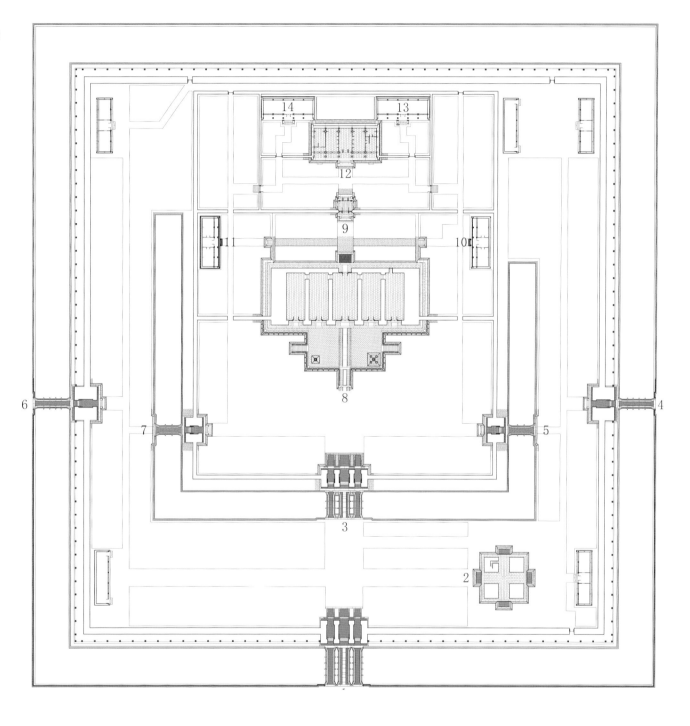

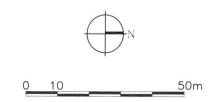

斋宫组群总平面图
Site plan of Complex of Palace of Abstinence

0　10　　　　　　50m

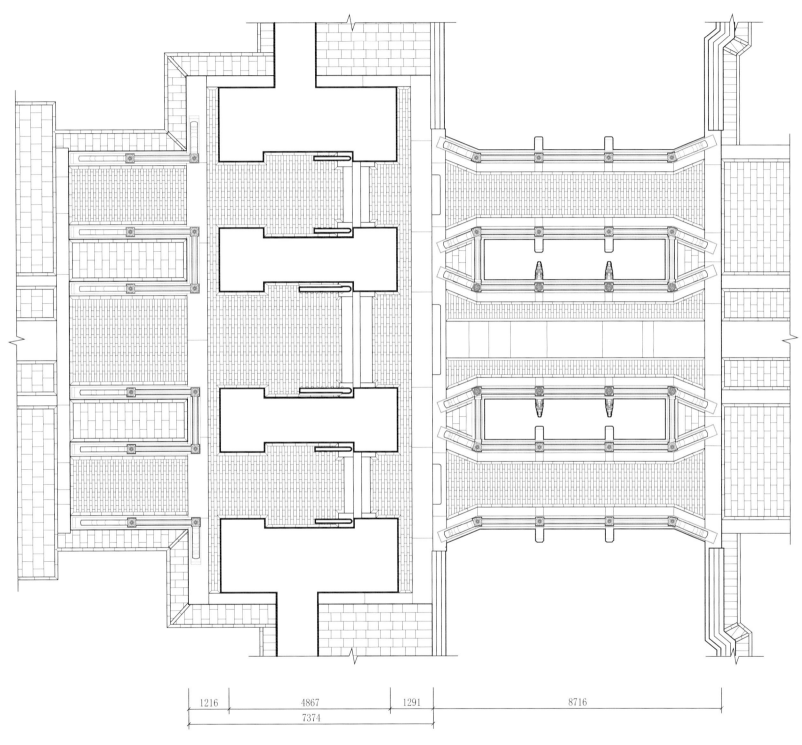

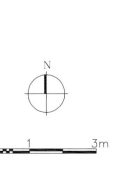

斋宫东内门及桥平面图
Plan of east inner gate and bridge of Palace of Abstinence

0 1 3m

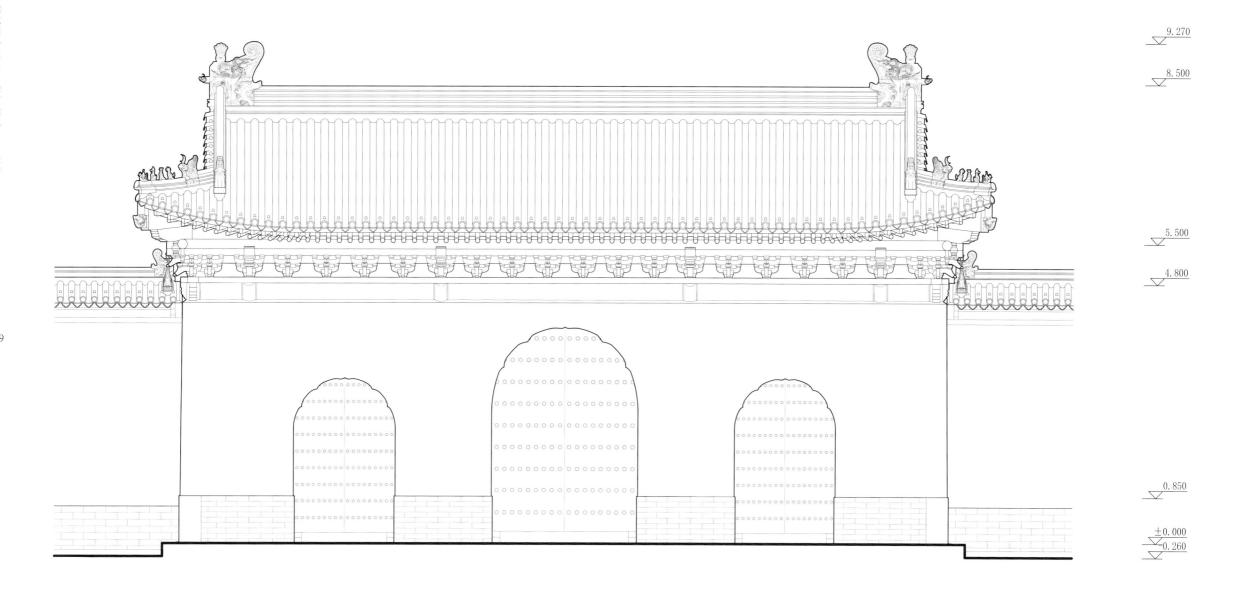

9.270

8.500

5.500

4.800

0.850

±0.000
-0.260

| 580 | 1067 | 3650 | 4780 | 3650 | 1067 | 580 |

15374

斋宫东内门正立面图
Front elevation of east inner gate of Palace of Abstinence

0　　　1　　　2m

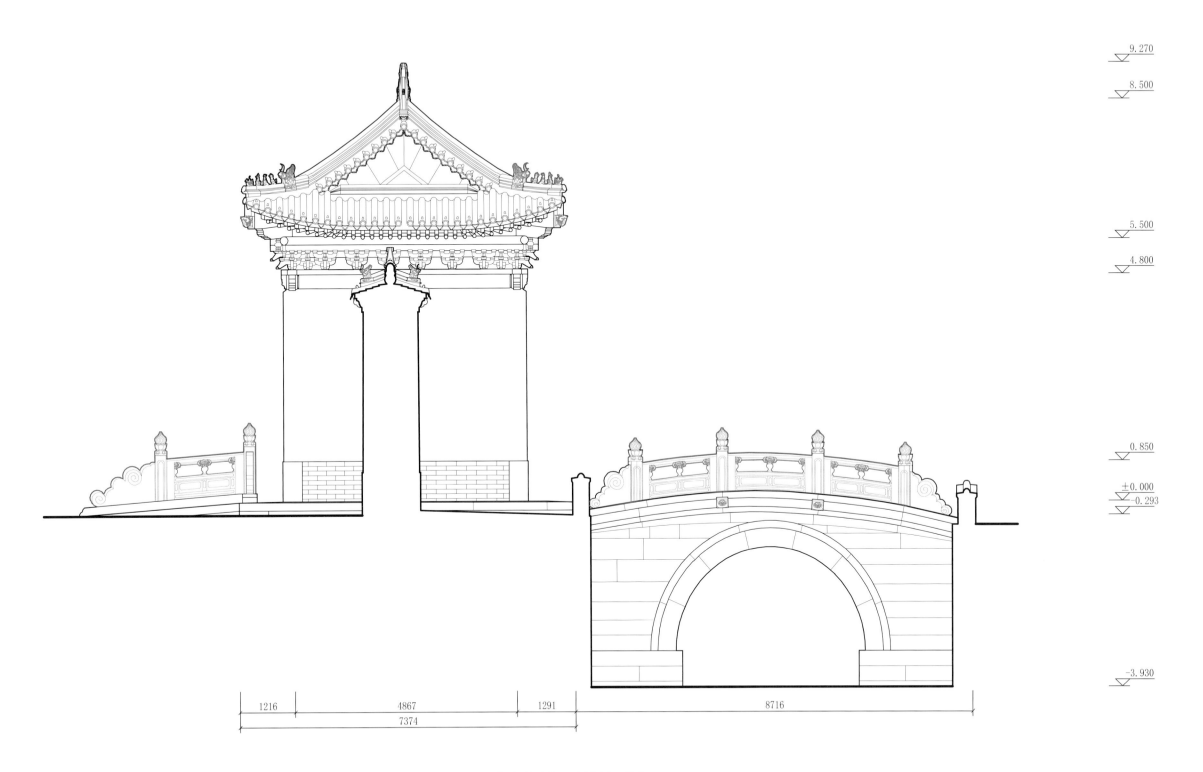

9.270

8.500

5.500

4.800

0.850

±0.000
−0.293

−3.930

1216   4867   1291   8716

7374

斋宫东内门及桥侧立面图
Side elevation of east inner gate and bridge of Palace of Abstinence

0   1   3m

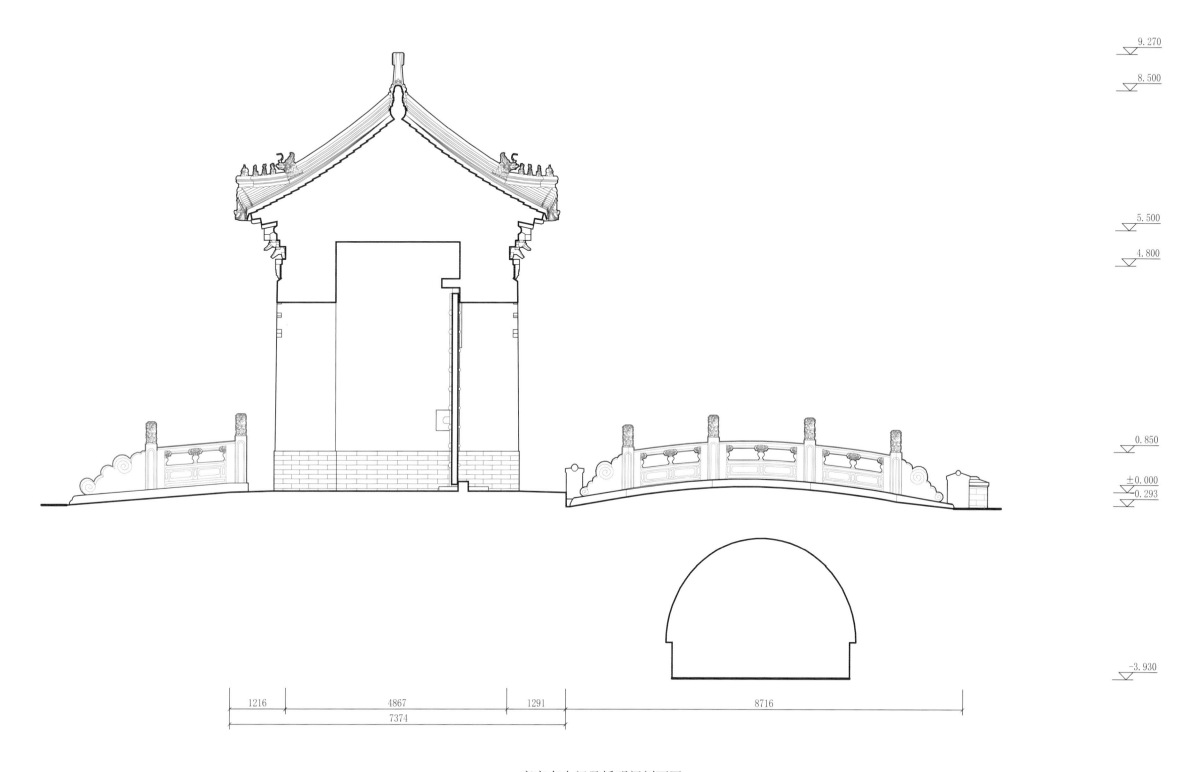

9.270

8.500

5.500

4.800

0.850

±0.000

-0.293

-3.930

1216　　　4867　　　1291　　　　　8716

7374

斋宫东内门及桥明间剖面图
Section of central-bay of east inner gate and bridge of Palace of Abstinence

0　　1　　　　3m

9.270

8.500

5.500

4.800

0.850

±0.000
-0.293

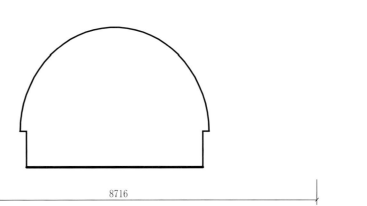

-3.930

1216   4867   1291   8716

7374

斋宫东内门及桥次间剖面图
Section of side-bay of east inner gate and bridge of Palace of Abstinence

0   1   3m

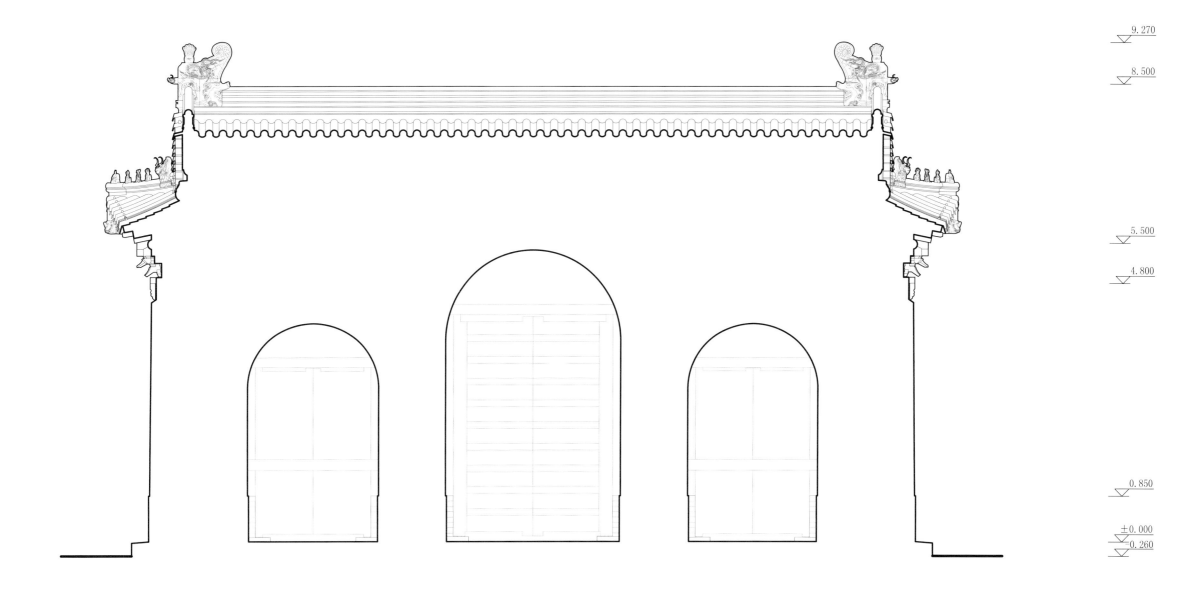

9.270

8.500

5.500

4.800

0.850

±0.000
0.260

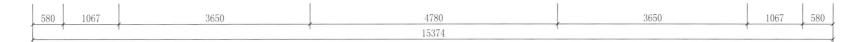

| 580 | 1067 | 3650 | 4780 | 3650 | 1067 | 580 |

15374

斋宫东内门纵剖面图
Longitudinal section of east inner gate of Palace of Abstinence

0　　1　　2m

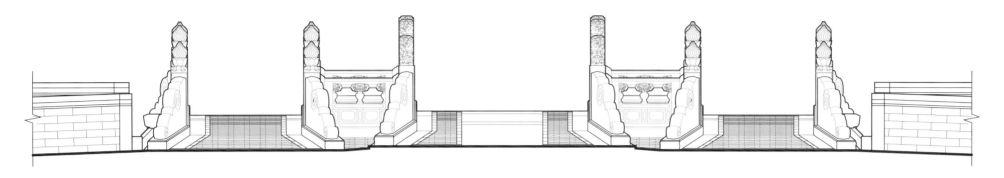

斋宫东内门桥正立面图
Front elevation of east inner gate and bridge of Palace of Abstinence

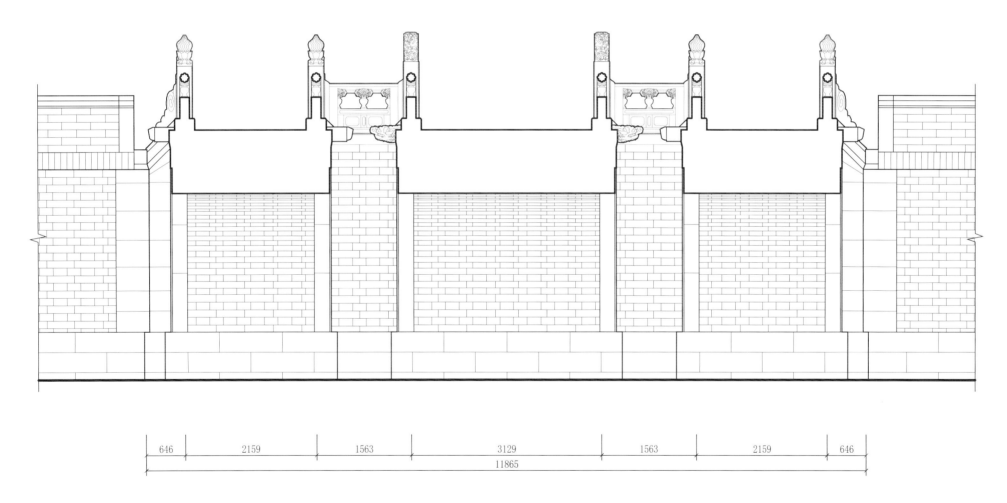

646	2159	1563	3129	1563	2159	646

11865

斋宫东内门桥纵剖面图
Longitudinal section of east inner gate and bridge of Palace of Abstinence

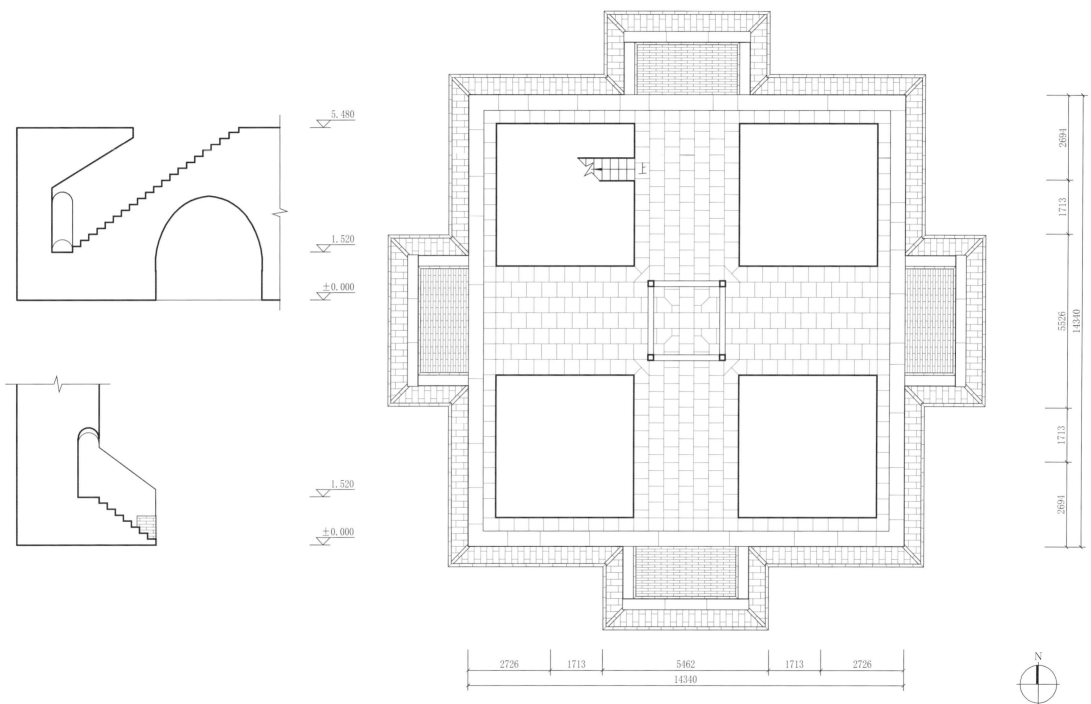

斋宫钟楼楼梯剖面图
Section of Bell Tower stairs of Palace of Abstinence

斋宫钟楼一层平面图
Plan of first floor of Bell Tower of Palace of Abstinence

0   1   3m

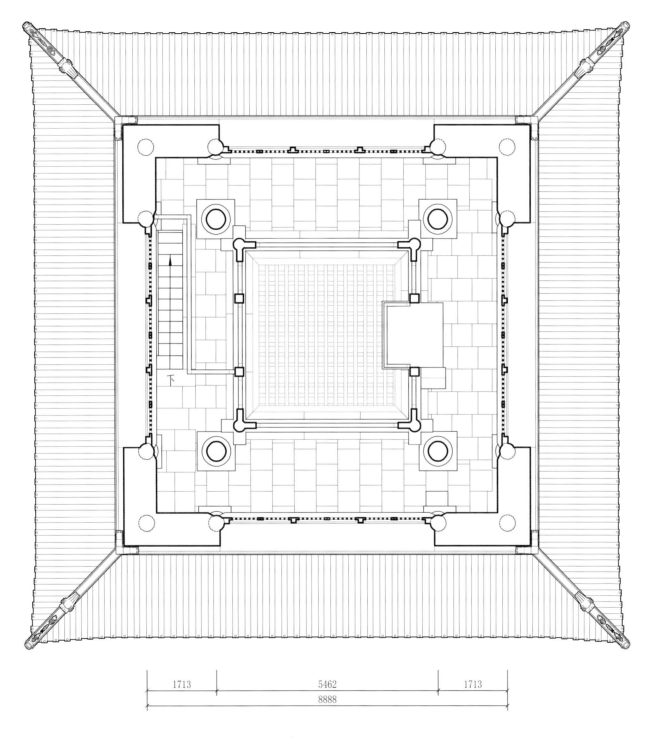

1713

5526

8952

1713

1713　　　5462　　　1713

8888

斋宫钟楼二层平面图

Plan of second floor of Bell Tower of Palace of Abstinence

0　　1　　　3m

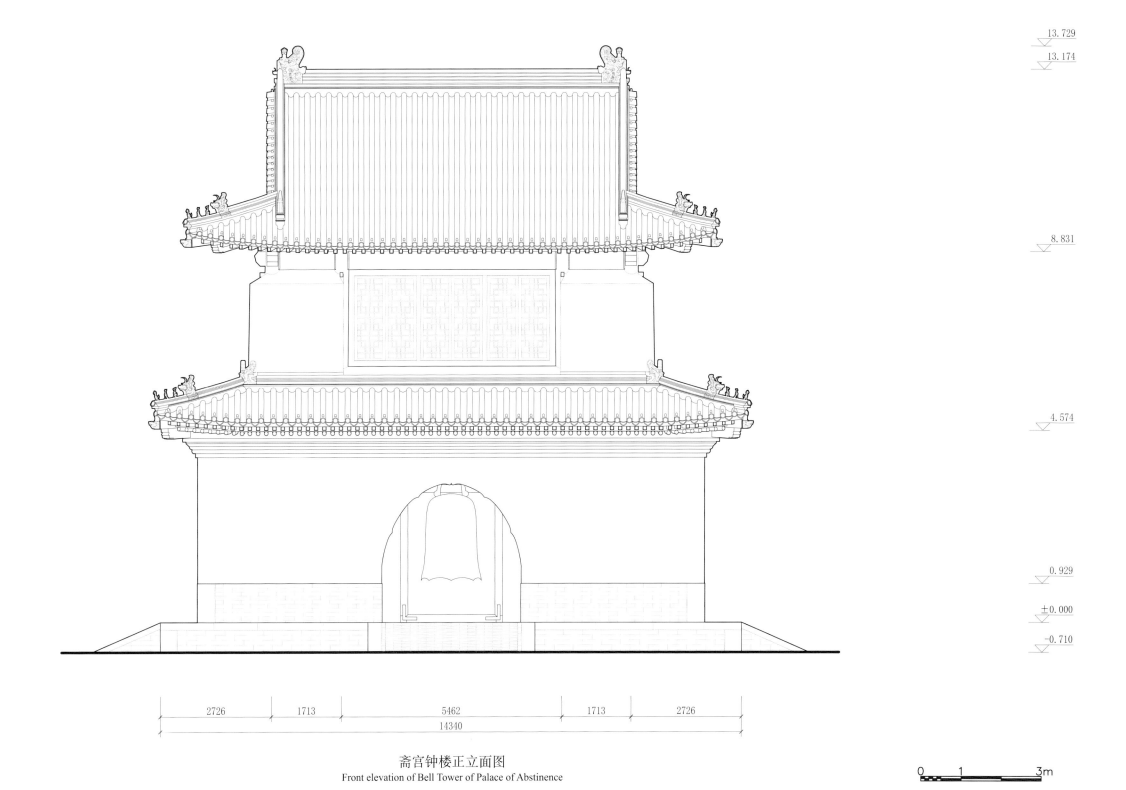

13.729

13.174

8.831

4.574

0.929

±0.000

−0.710

2726 | 1713 | 5462 | 1713 | 2726

14340

斋宫钟楼正立面图
Front elevation of Bell Tower of Palace of Abstinence

0   1   3m

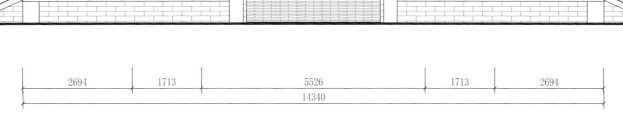

13.729
13.174

8.831

4.574

0.929

±0.000

-0.710

| 2694 | 1713 | 5526 | 1713 | 2694 |

14340

斋宫钟楼侧立面图
Side elevation of Bell Tower of Palace of Abstinence

0 1 3m

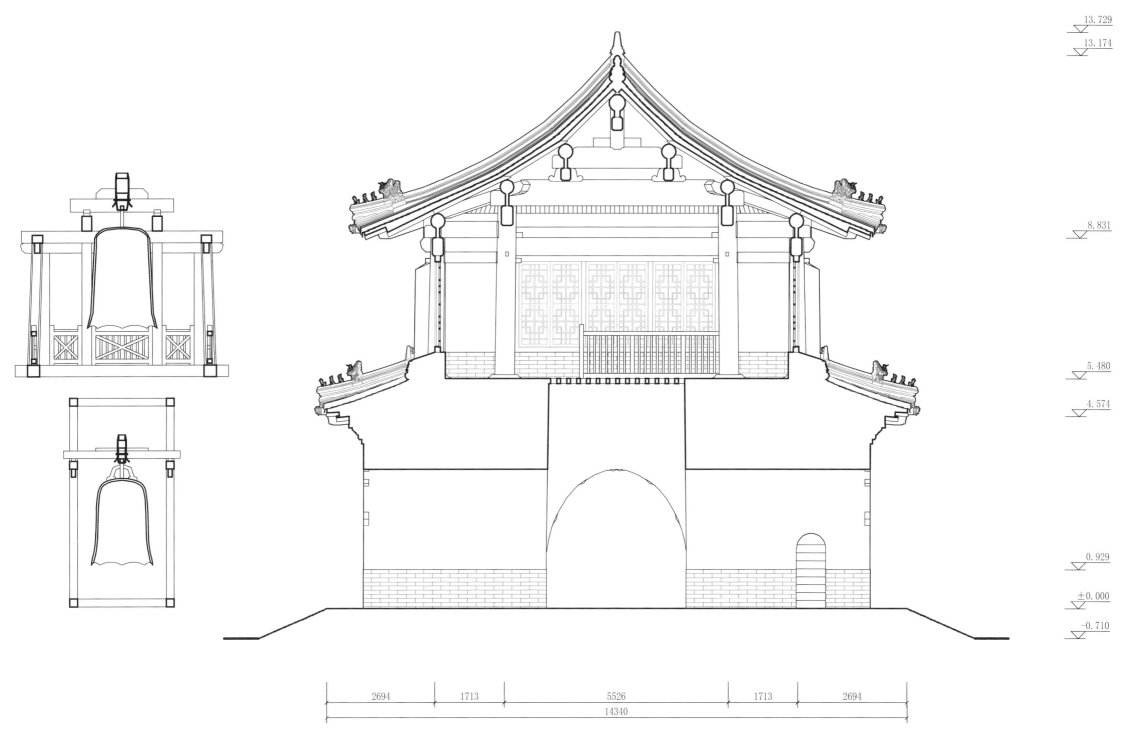

13.729

13.174

8.831

5.480

4.574

0.929

±0.000

-0.710

| 2694 | 1713 | 5526 | 1713 | 2694 |

14340

钟与钟架横剖面图
Cross-section of bell and frame

斋宫钟楼横剖面图
Cross-section of Bell Tower of Palace of Abstinence

0 1 3m

13.729

13.174

8.831

5.480

4.574

0.929

±0.000

-0.710

| 2726 | 1713 | 5462 | 1713 | 2726 |

14340

钟与钟架纵剖面图
Longitudinal section of bell and frame

斋宫钟楼纵剖面图
Longitudinal section of Bell Tower of Palace of Abstinence

0 1 3m

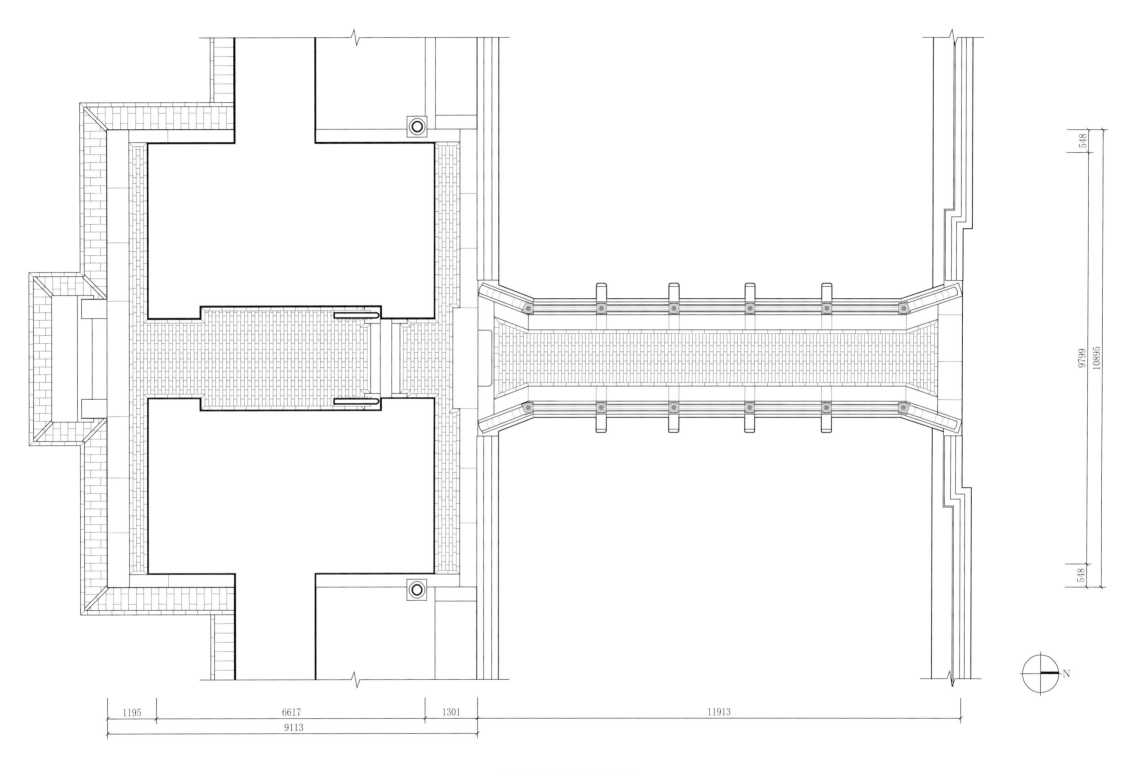

斋宫北外门及桥平面图
Plan of north outer gate and bridge of Palace of Abstinence

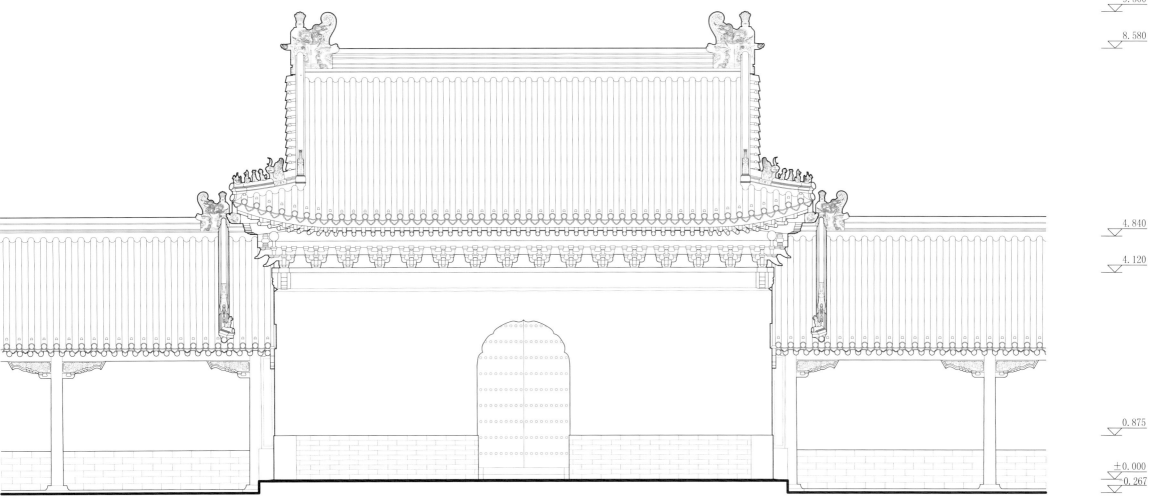

9.300

8.580

4.840

4.120

0.875

±0.000
0.267

548 | 9799 | 548
10895

斋宫北外门正立面图
Front elevation of north outer gate of Palace of Abstinence

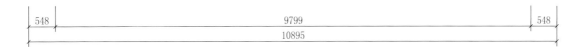

0   1   3m

143

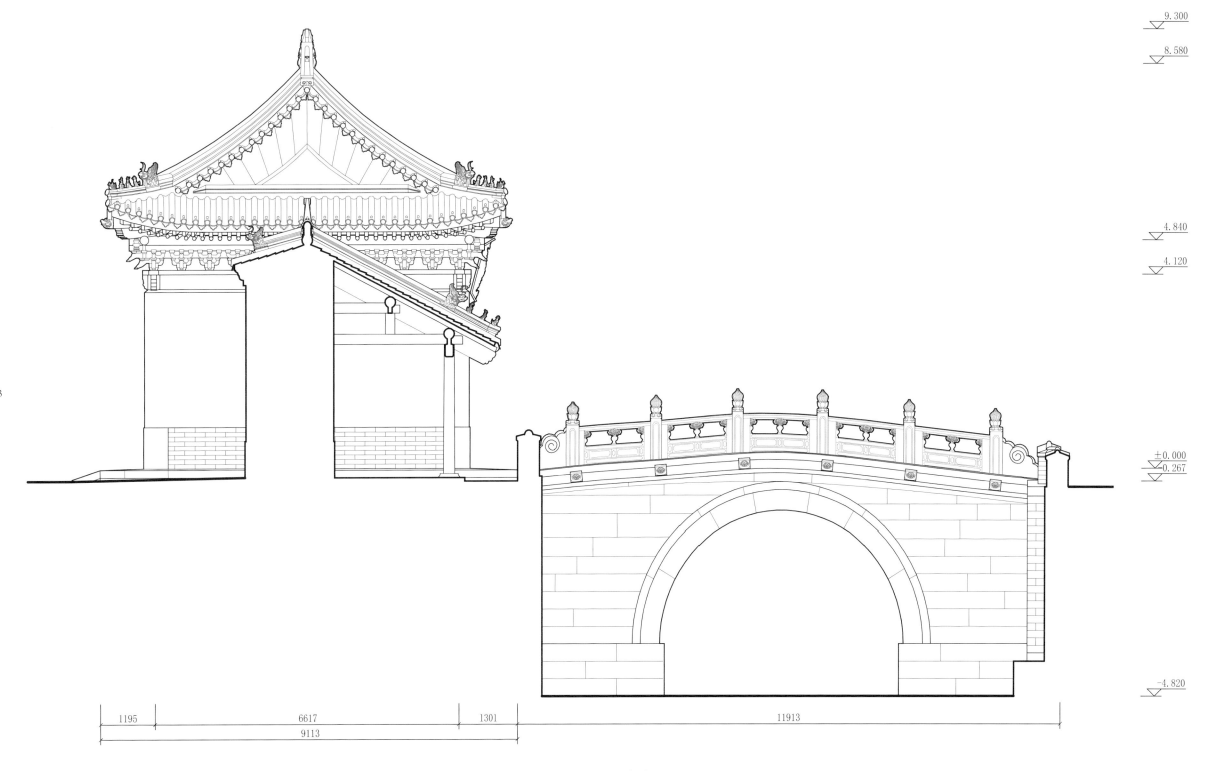

9.300

8.580

4.840

4.120

±0.000
-0.267

-4.820

1195　6617　1301　11913

9113

斋宫北外门及桥侧立面图
Side elevation of north outer gate and bridge of Palace of Abstinence

0　1　3m

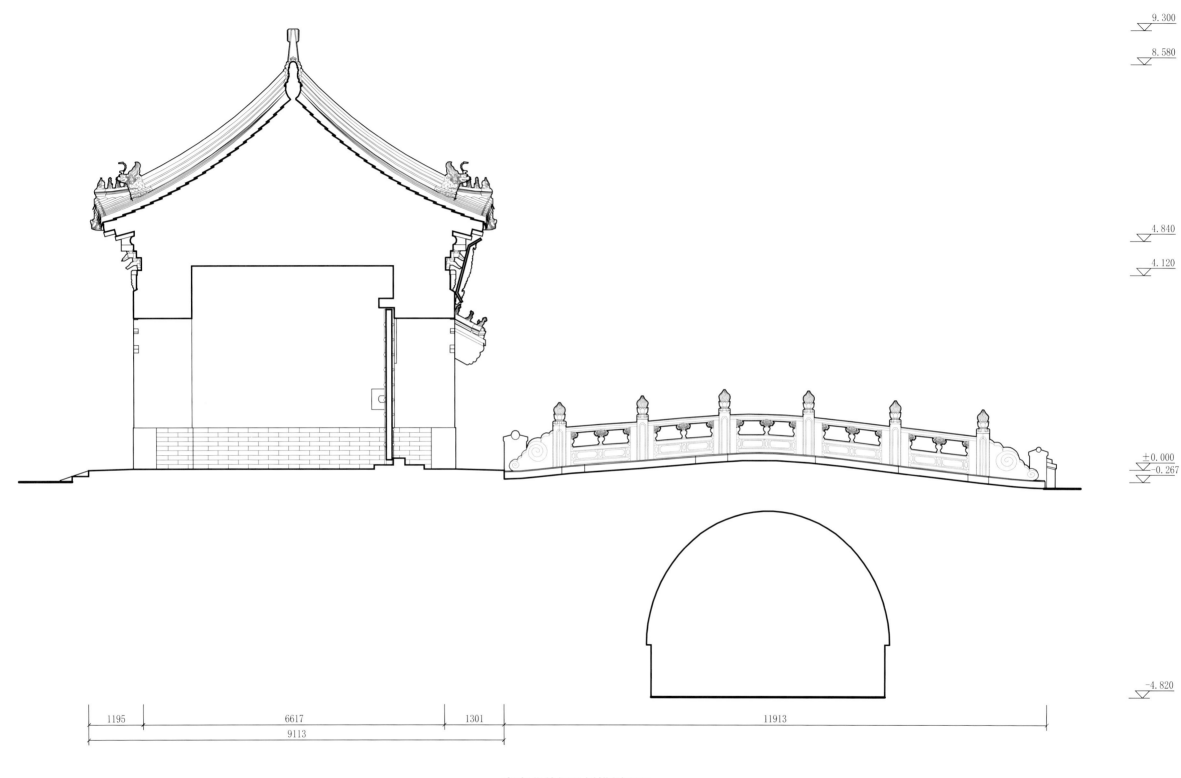

9.300

8.580

4.840

4.120

±0.000
-0.267

-4.820

1195	6617	1301	11913

9113

斋宫北外门及桥横剖面图
Cross-section of north outer gate and bridge of Palace of Abstinence

0 1 3m

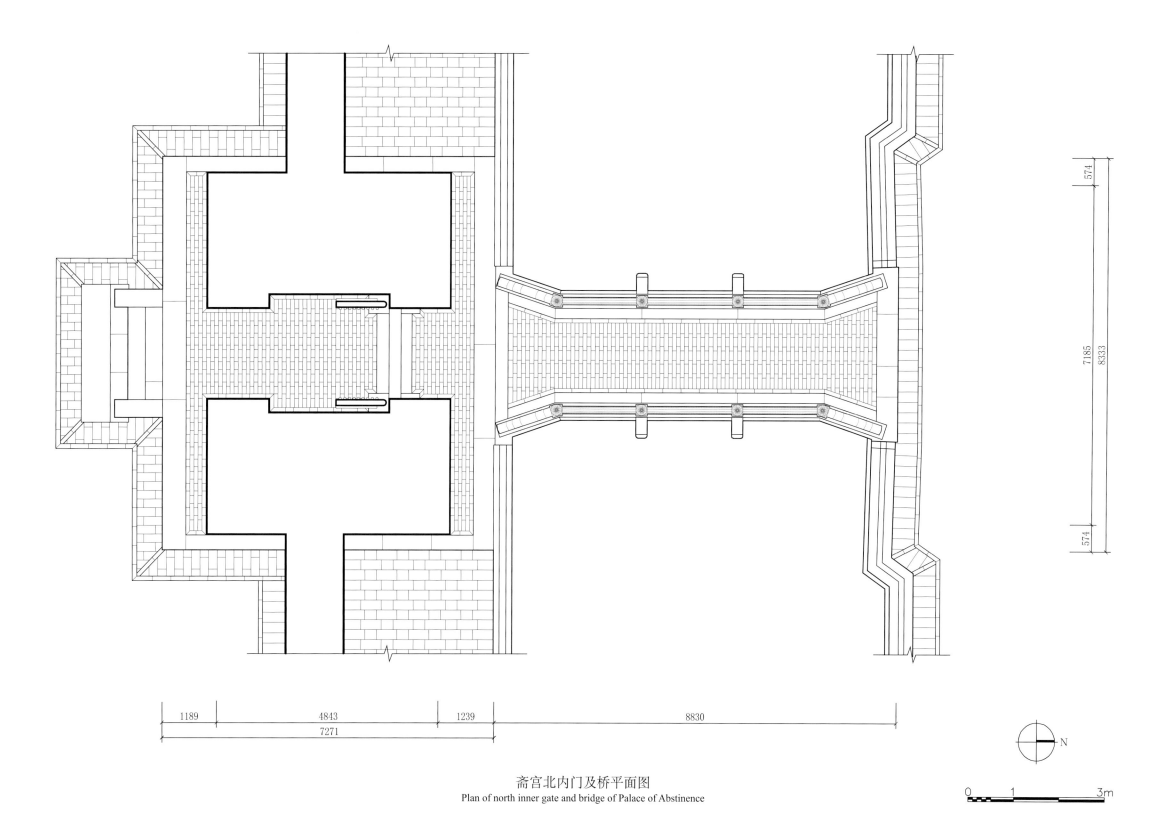

574

7185

8333

574

1189   4843   1239

7271

8830

斋宫北内门及桥平面图
Plan of north inner gate and bridge of Palace of Abstinence

N

0   1   3m

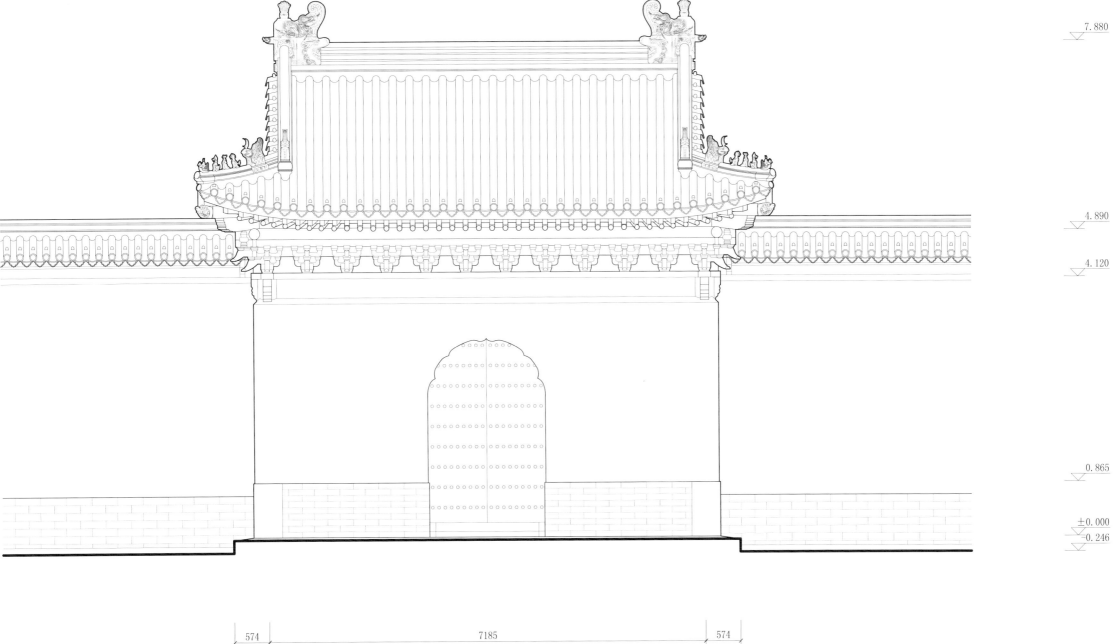

8.540

7.880

4.890

4.120

0.865

±0.000
-0.246

574   7185   574

8333

斋宫北内门正立面图

Front elevation of north inner gate of Palace of Abstinence

0   1   2m

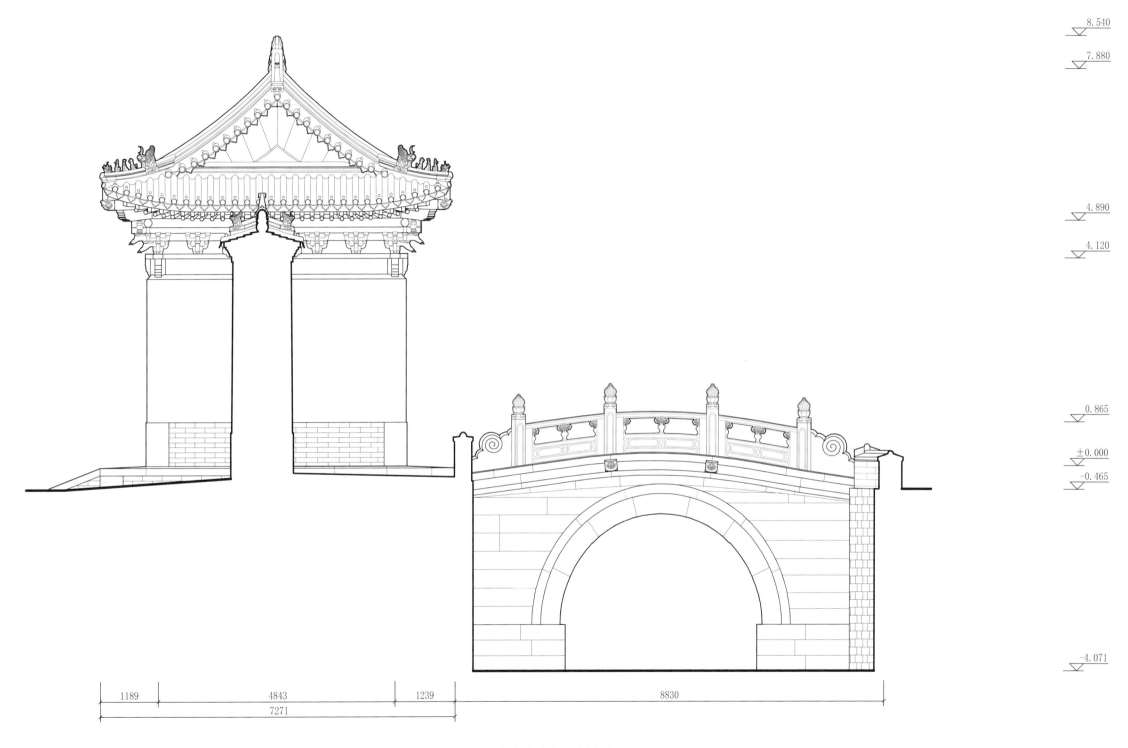

8.540

7.880

4.890

4.120

0.865

±0.000

-0.465

-4.071

1189   4843   1239   8830

7271

斋宫北内门及桥侧立面图
Side elevation of north inner gate and bridge of Palace of Abstinence

0   1   3m

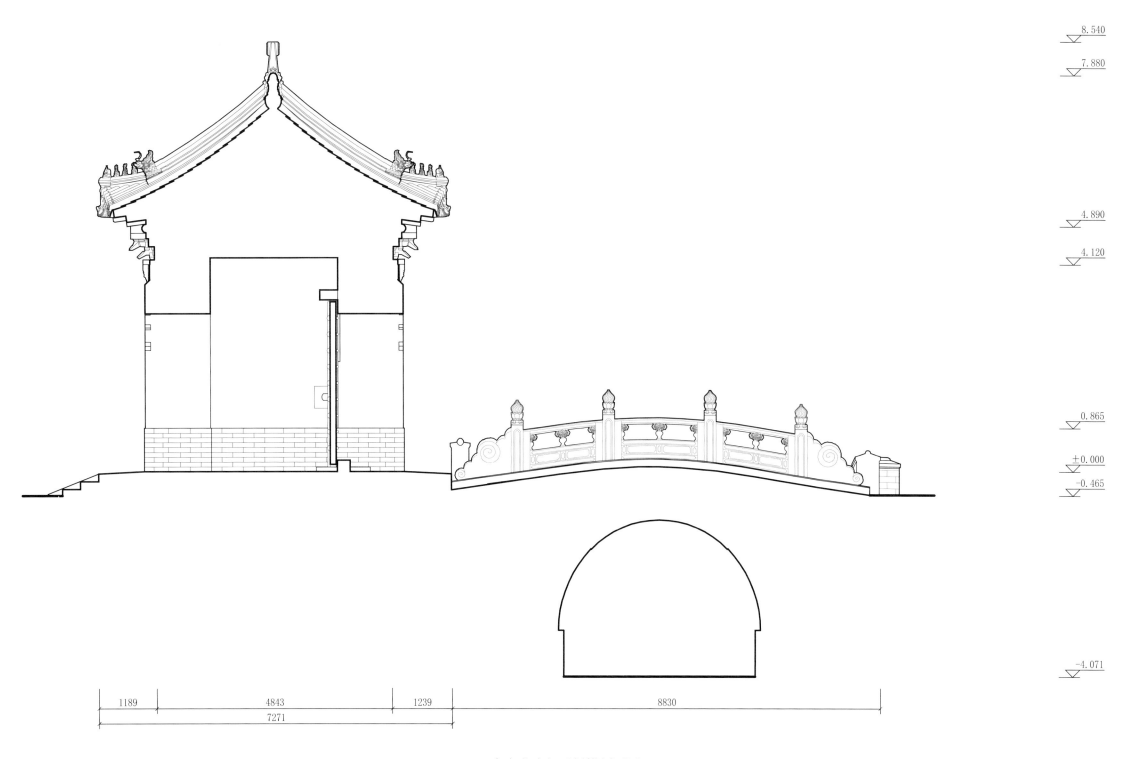

8.540

7.880

4.890

4.120

0.865

±0.000

-0.465

-4.071

1189   4843   1239   8830

7271

斋宫北内门及桥横剖面图
Cross-section of north inner gate and bridge of Palace of Abstinence

0   1   3m

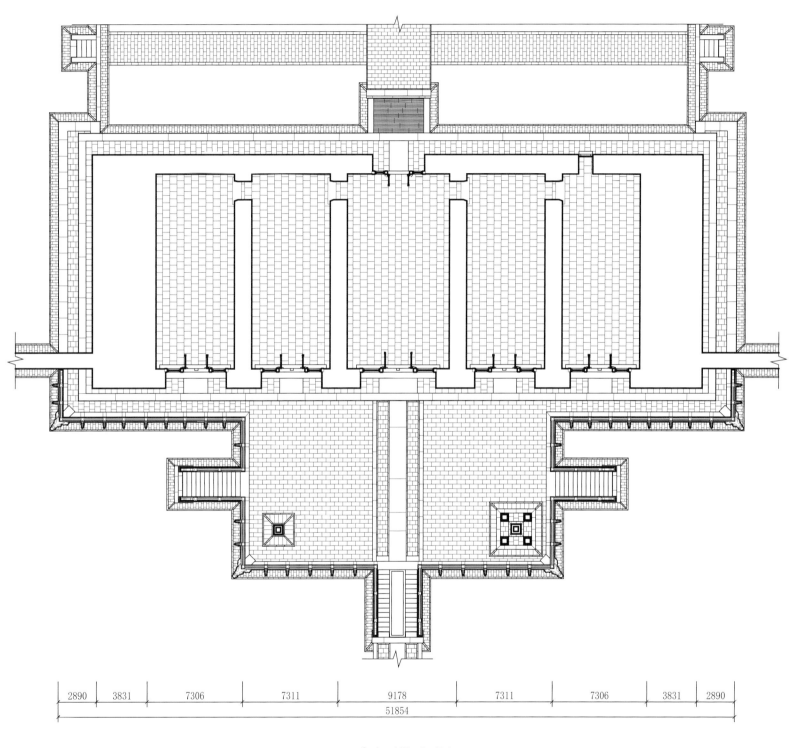

2890　3831　7306　7311　9178　7311　7306　3831　2890

51854

斋宫正殿平面图
Plan of main hall of Palace of Abstinence

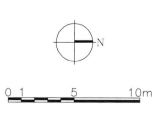

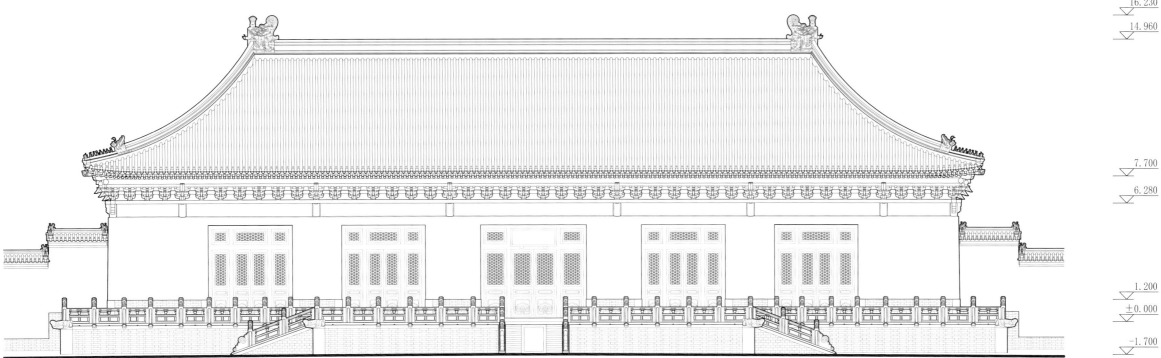

16.230

14.960

7.700

6.280

150

1.200

±0.000

-1.700

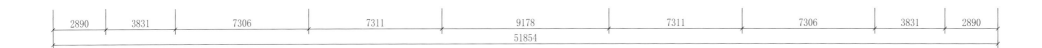

| 2890 | 3831 | 7306 | 7311 | 9178 | 7311 | 7306 | 3831 | 2890 |

51854

斋宫正殿正立面图
Front elevation of main hall of Palace of Abstinence

0 1 5m

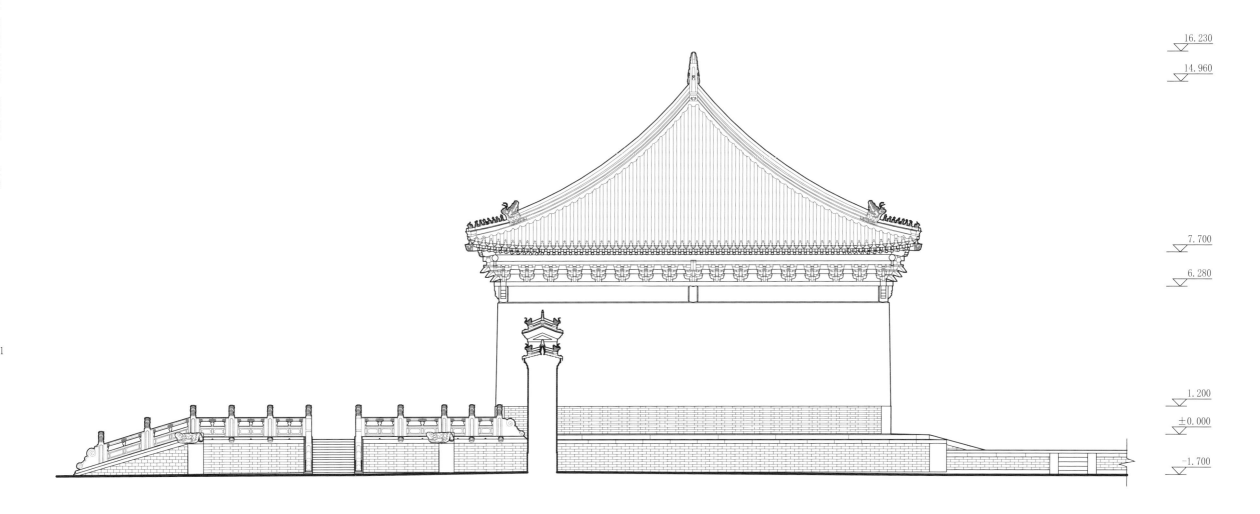

16.230

14.960

7.700

6.280

1.200

±0.000

-1.700

| 13769 | 8326 | 8326 | 1904 | 897 |

33222

斋宫正殿侧立面图
Side elevation of main hall of Palace of Abstinence

0   1   5m

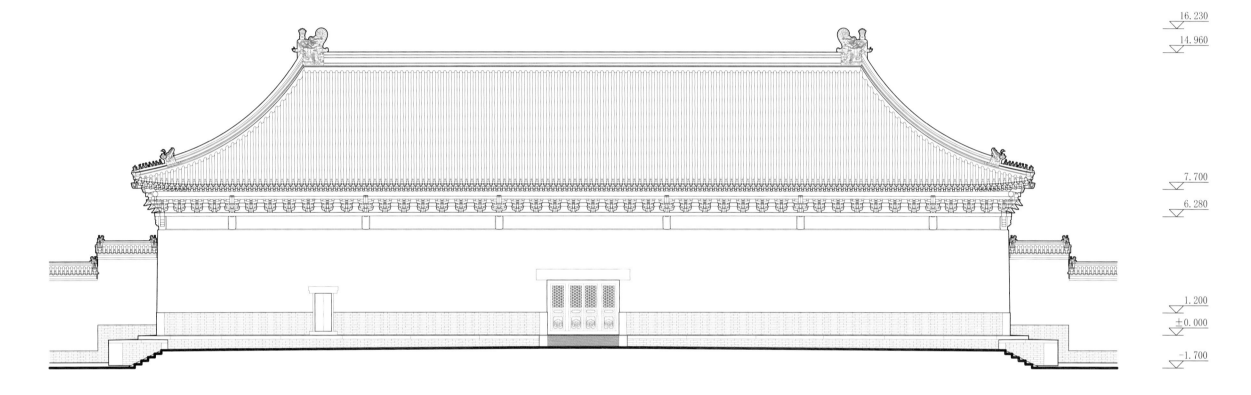

16.230
14.960

7.700
6.280

1.200
±0.000

-1.700

| 2890 | 3831 | 7306 | 7311 | 9178 | 7311 | 7306 | 3831 | 2890 |

51854

斋宫正殿背立面图
Rear elevation of main hall of Palace of Abstinence

0 1 5m

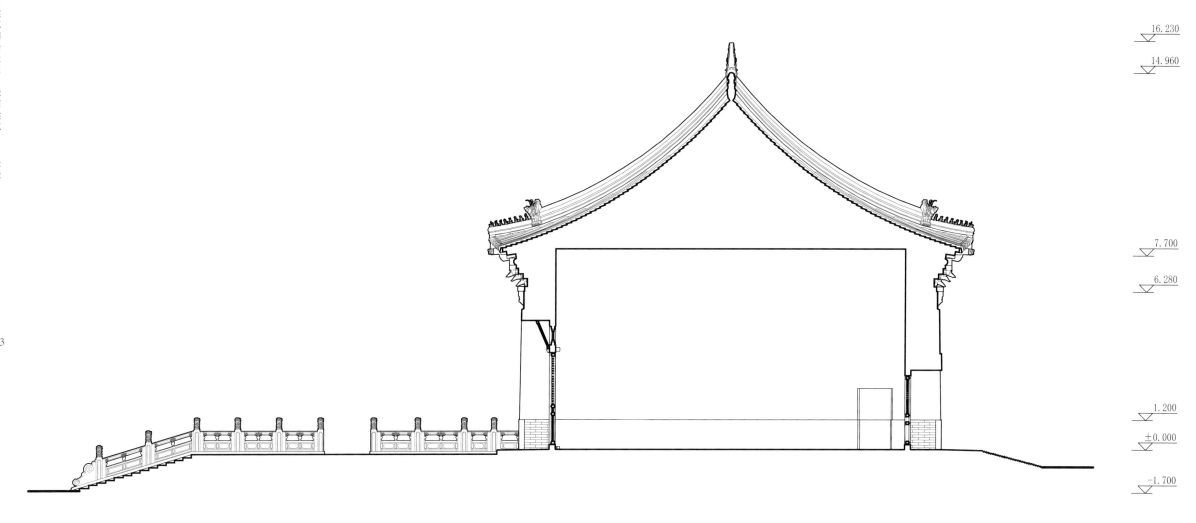

16.230

14.960

7.700

6.280

1.200

±0.000

-1.700

| 12507 | 1262 | 8326 | 8326 | 1904 | 897 |

20715

斋宫正殿明间剖面图

Section of central-bay of main hall of Palace of Abstinence

0   1   5m

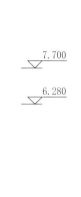

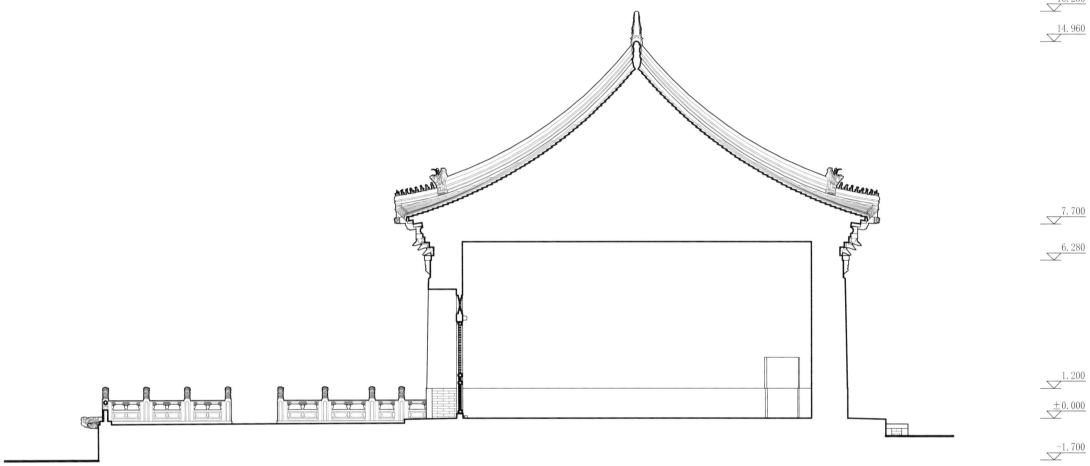

16.230

14.960

7.700

6.280

1.200

±0.000

-1.700

1262　8326　8326　1904　897

20715

斋宫正殿次间剖面图
Section of side-bay of main hall of Palace of Abstinence

0　1　5m

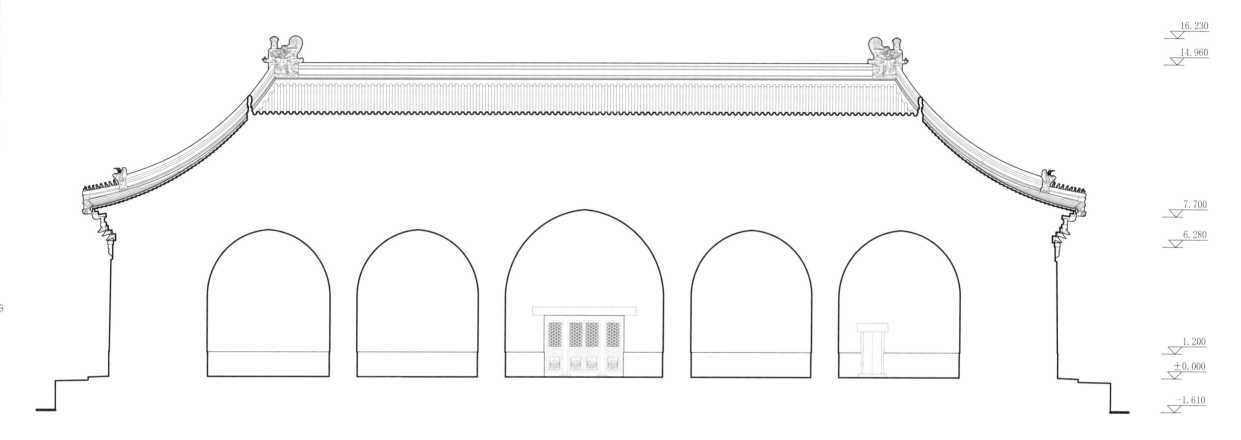

16.230

14.960

7.700

6.280

1.200

±0.000

-1.610

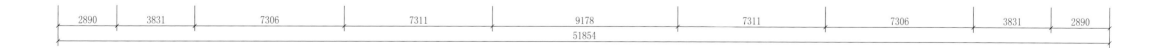

2890 3831 7306 7311 9178 7311 7306 3831 2890

51854

斋宫正殿纵剖面图
Longitudinal section of main hall of Palace of Abstinence

0   1   5m

斋宫正殿丹陛大样图
Danbi of main hall of Palace of Abstinence

0　0.1　0.5m

斋宫正殿铜人亭正立面图
Front elevation of Pavilion of the Bronze
Statue from main hall of Palace of Abstinence

斋宫正殿铜人亭剖面图
Section of Pavilion of the Bronze Statue
from main hall of Palace of Abstinence

斋宫正殿铜人亭大样图
Pavilion of the Bronze Statue from main hall of Palace of Abstinence

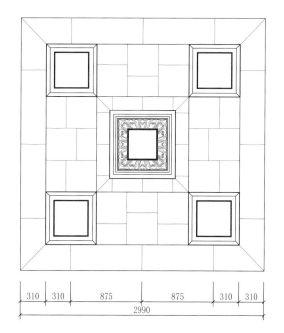

斋宫正殿铜人亭平面图
Plan of Pavilion of the Bronze Statue from main hall of Palace of Abstinence

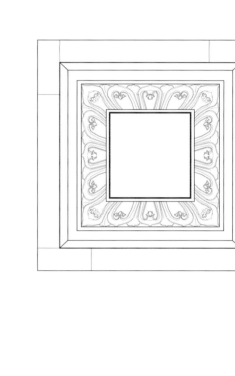

2.198

1.800

1.018

0.470

±0.000

218 352 218

788

0 0.1   0.5m

斋宫正殿时辰亭侧立面图
Side elevation of Time Pavilion from main
hall of Palace of Abstinence

斋宫正殿时辰亭正立面图
Front elevation of Time Pavilion from main
hall of Palace of Abstinence

斋宫正殿时辰亭背立面图
Rear elevation of Time Pavilion from main
hall of Palace of Abstinence

斋宫正殿时辰亭平面图
Plan of Time Pavilion from main hall of
Palace of Abstinence

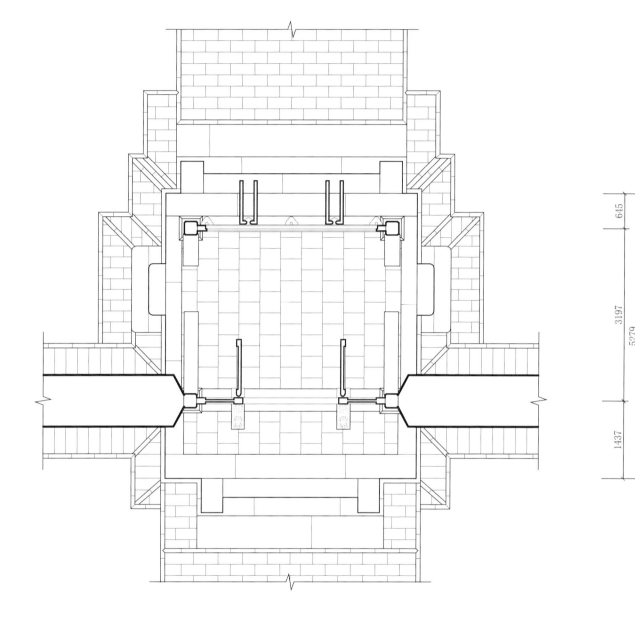

斋宫垂花门抱鼓石大样图
Drum stone of *chuihuamen* **of Palace of Abstinence**

0   0.1   0.3m

斋宫垂花门平面图
Plan of *chuihuamen* **of Palace of Abstinence**

0   1   2m

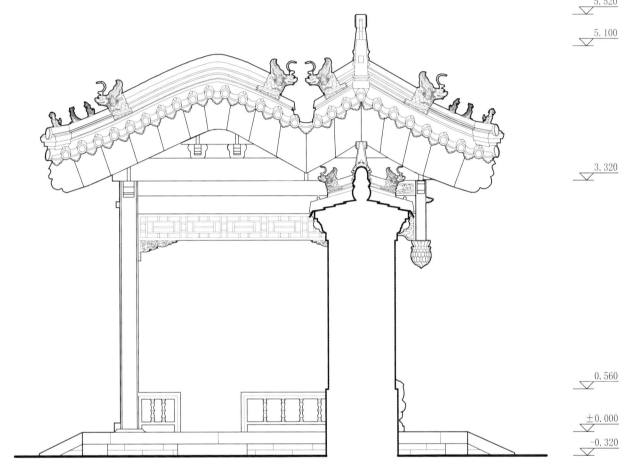

5.520

5.100

3.320

0.560

±0.000

-0.320

479 3862 479
4820

645 3197 1437
5279

斋宫垂花门正立面图
Front elevation of *chuihuamen* of Palace of Abstinence

斋宫垂花门侧立面图
Side elevation of *chuihuamen* of Palace of Abstinence

0   0.5   1m

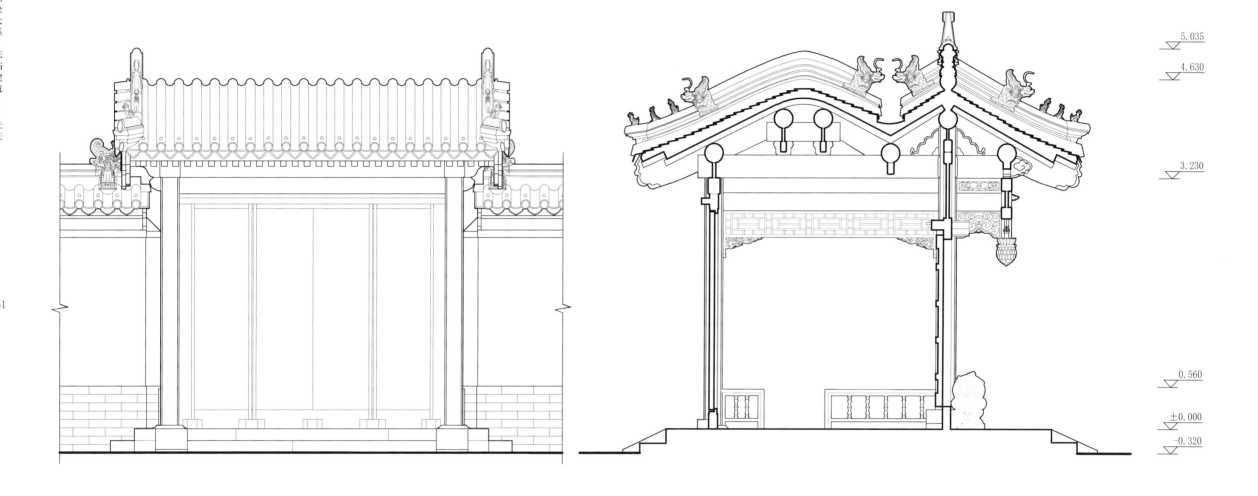

5.035

4.630

3.230

0.560

±0.000

-0.320

479   3862   479

4820

645   3197   1437

5279

斋宫垂花门背立面图
Rear elevation of *chuihuamen* of Palace of Abstinence

斋宫垂花门横剖面图
Cross-section of *chuihuamen* of Palace of Abstinence

0   0.5   1m

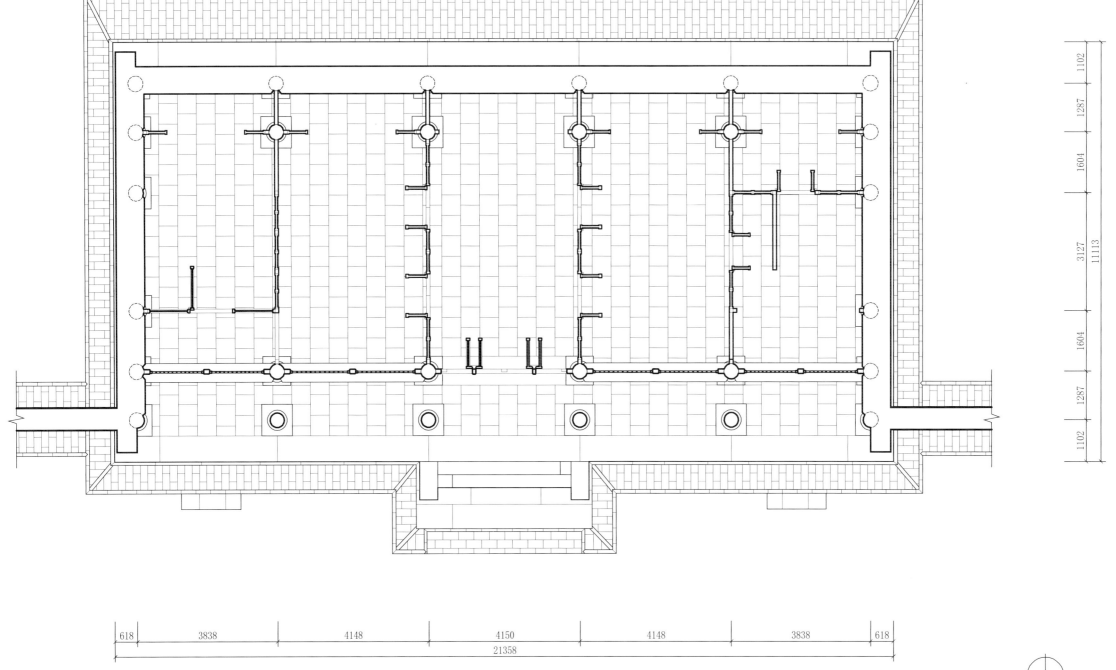

1102 1287 1604 3127 11113 1604 1287 1102

618 3838 4148 4150 4148 3838 618

21358

斋宫寝殿平面图
Plan of sleeping hall of Palace of Abstinence

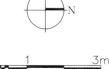

N

0 1 3m

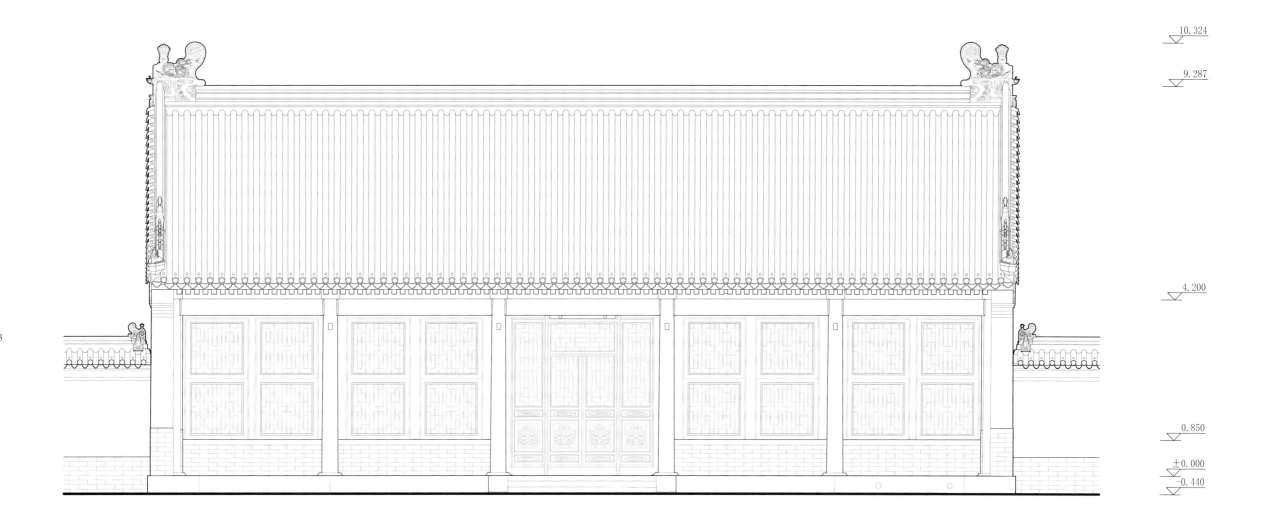

618 | 3838 | 4148 | 4150 | 4148 | 3838 | 618

21358

斋宫寝殿正立面图
Front elevation of sleeping hall of Palace of Abstinence

0 1 3m

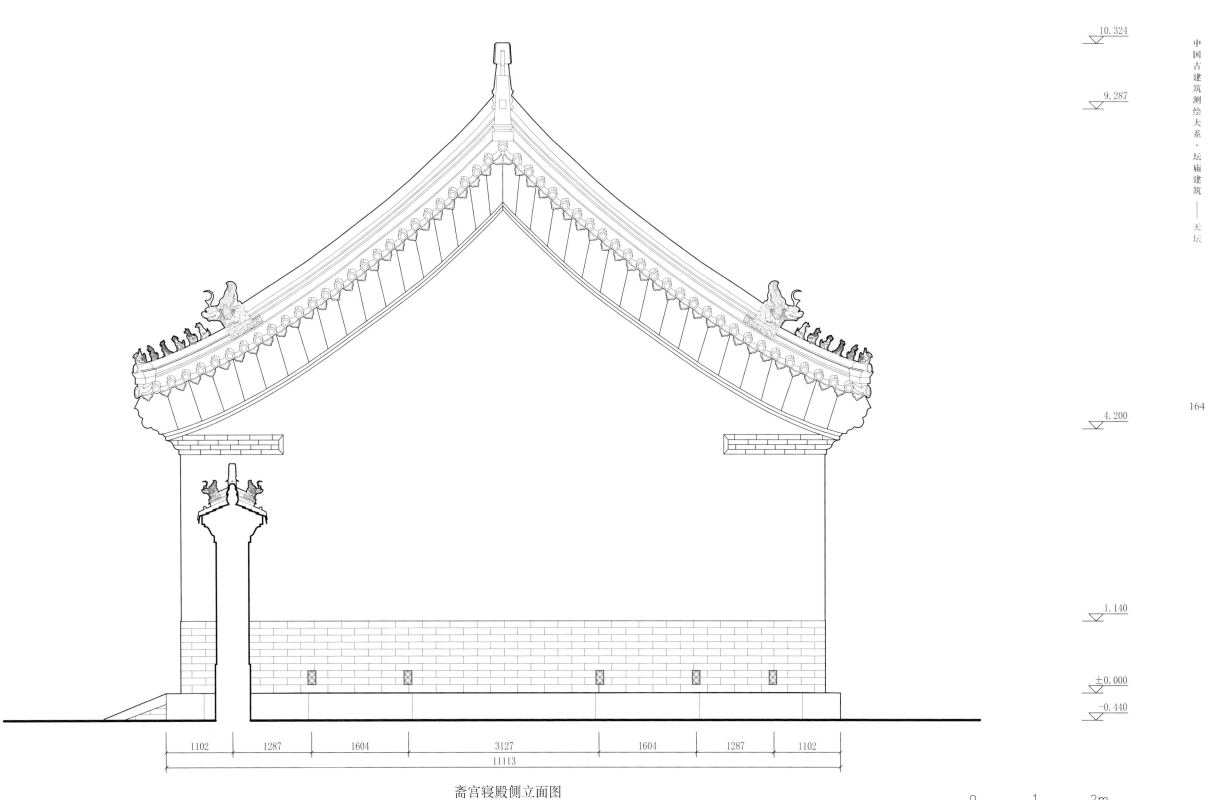

10.324

9.287

4.200

1.140

±0.000

-0.440

| 1102 | 1287 | 1604 | 3127 | 1604 | 1287 | 1102 |

11113

斋宫寝殿侧立面图
Side elevation of sleeping hall of Palace of Abstinence

0　　　1　　　2m

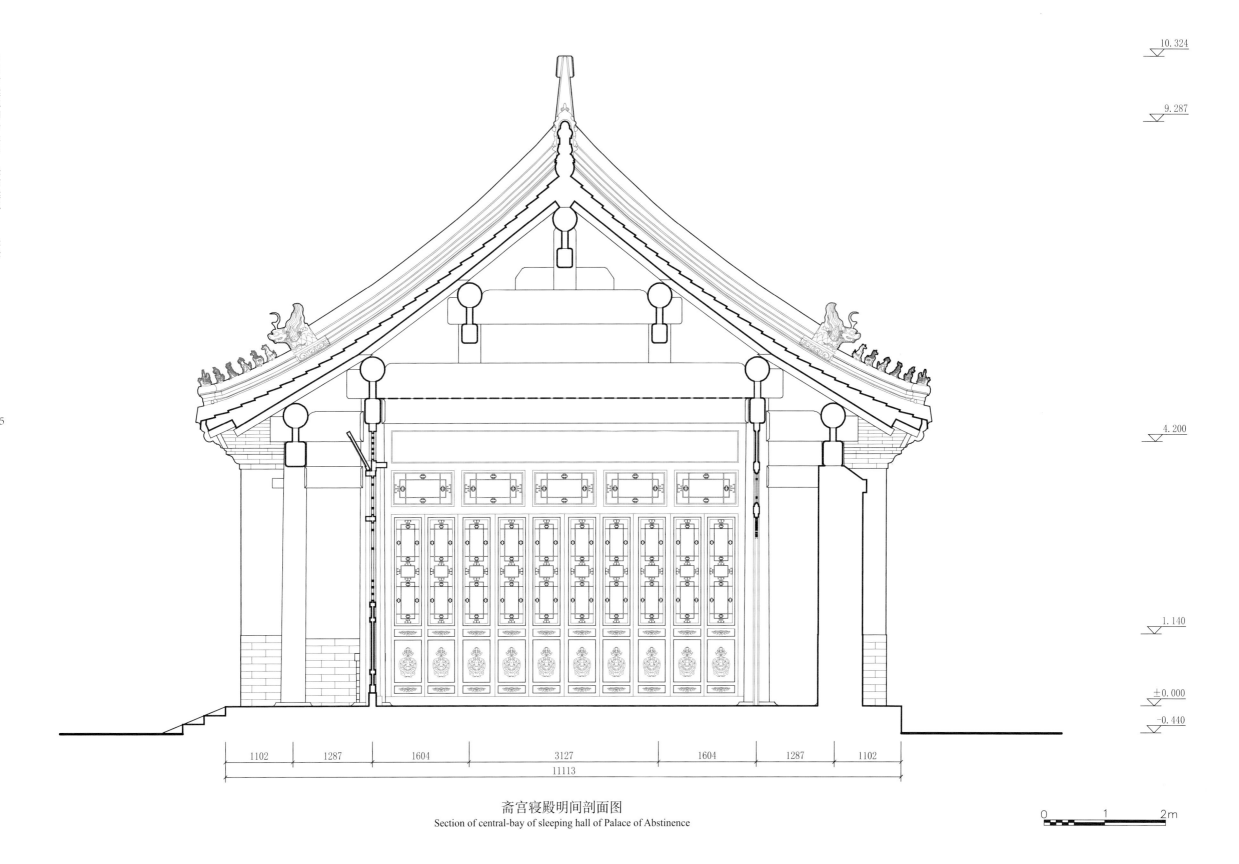

10. 324

9. 287

4. 200

1. 140

±0. 000

-0. 440

| 1102 | 1287 | 1604 | 3127 | 1604 | 1287 | 1102 |

11113

斋宫寝殿明间剖面图
Section of central-bay of sleeping hall of Palace of Abstinence

0   1   2m

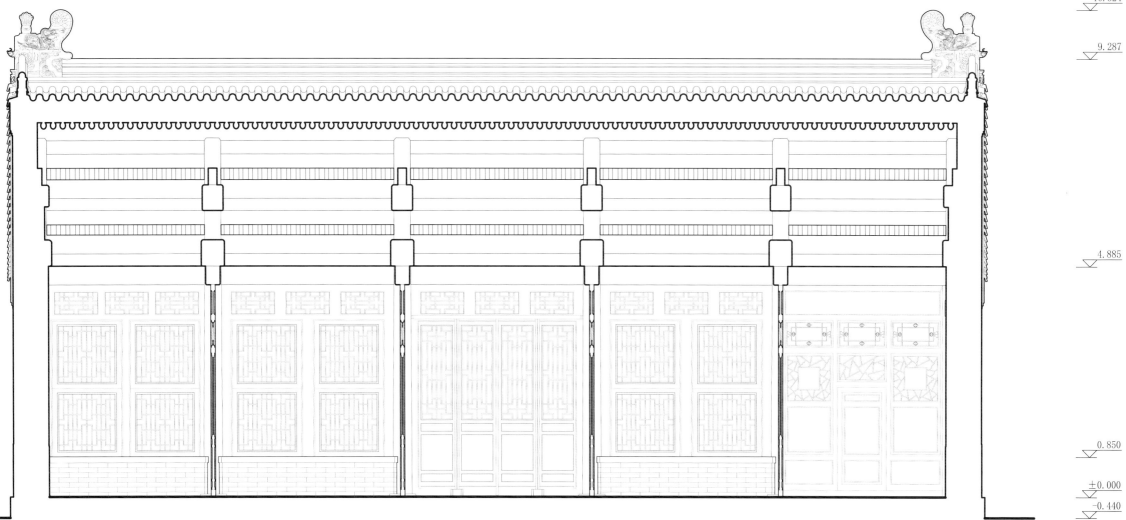

10.324

9.287

4.885

166

0.850

±0.000

-0.440

618　3838　4148　4150　4148　3838　618

21358

斋宫寝殿纵剖面图
Longitudinal section of sleeping hall of Palace of Abstinence

0　1　2m

神乐署组群
Complex of Divine Music Administration

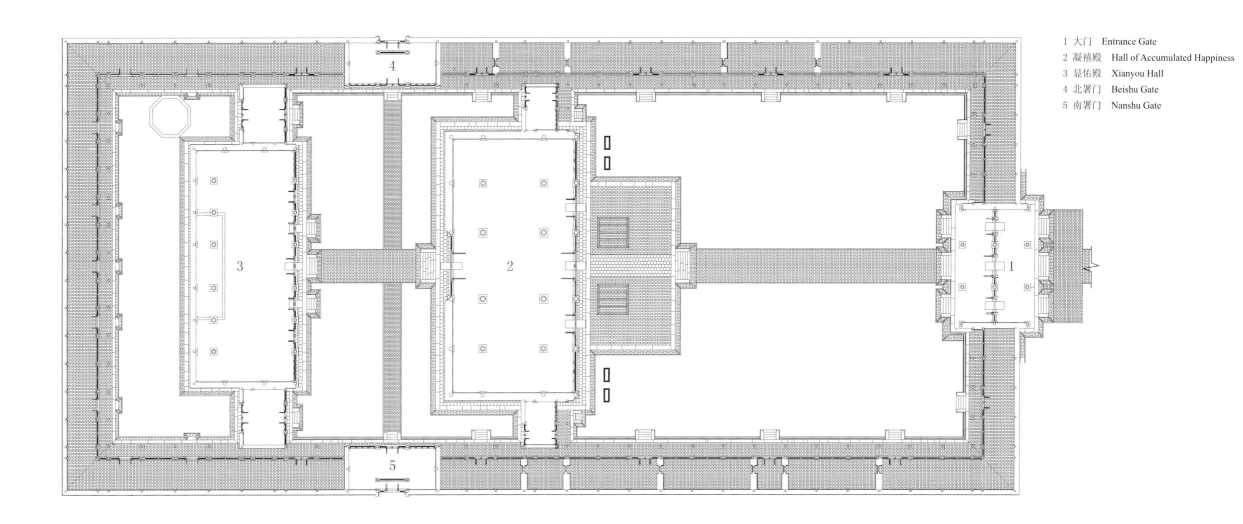

1　大门　Entrance Gate
2　凝禧殿　Hall of Accumulated Happiness
3　显佑殿　Xianyou Hall
4　北署门　Beishu Gate
5　南署门　Nanshu Gate

神乐署组群总平面图
Site plan of Complex of Divine Music Administration

0　5　10m

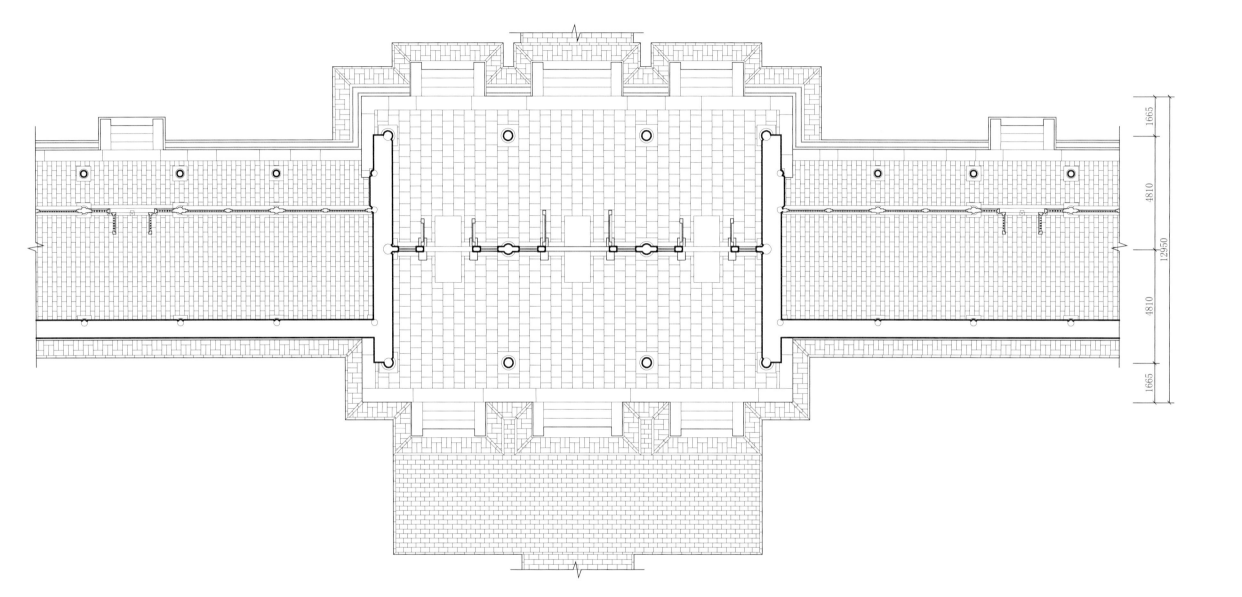

神乐署大门平面图
Plan of entrance gate of Divine Music Administration

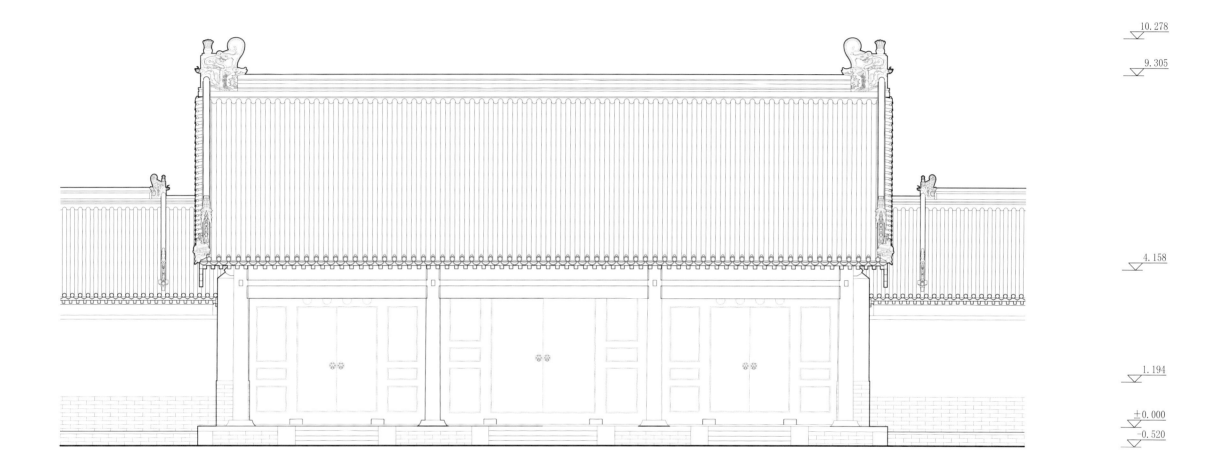

10. 278

9. 305

4. 158

1. 194

±0. 000

-0. 520

| 1160 | 5240 | 6080 | 5240 | 1160 |

18880

神乐署大门正立面图
Front elevation of entrance gate of Divine Music Administration

0　　1　　　　3m

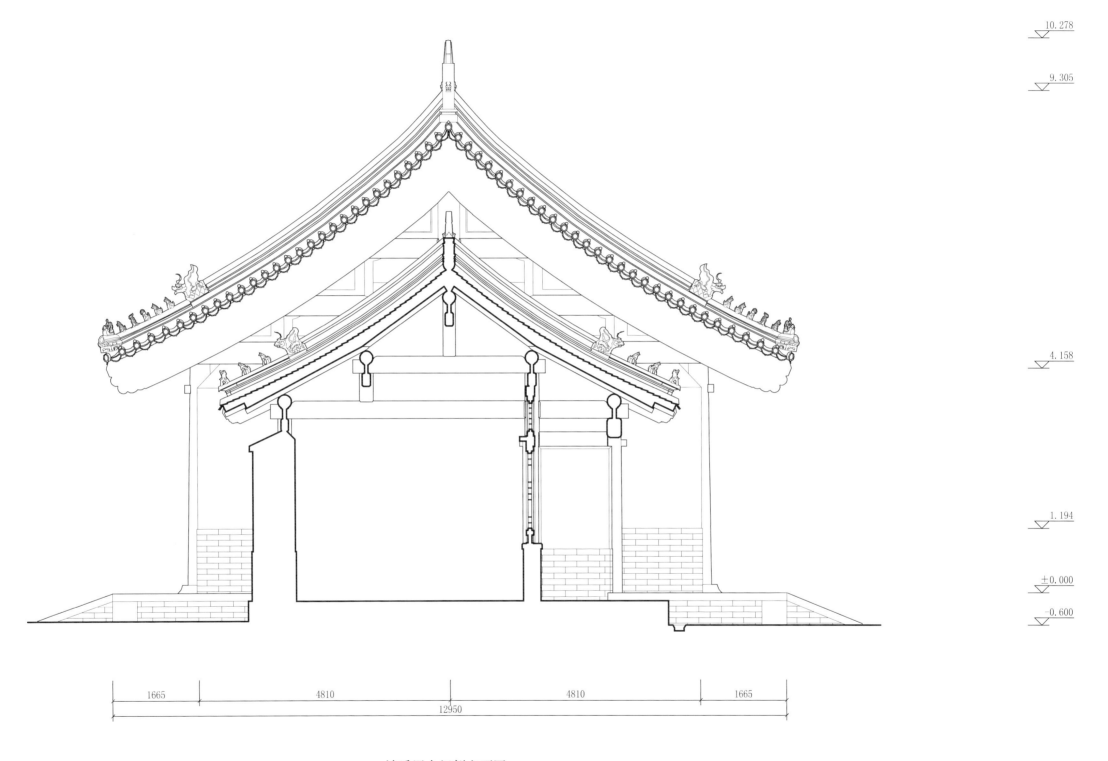

10.278

9.305

4.158

1.194

±0.000

-0.600

1665   4810   4810   1665

12950

神乐署大门侧立面图

Side elevation of entrance gate of Divine Music Administration

0   1   2m

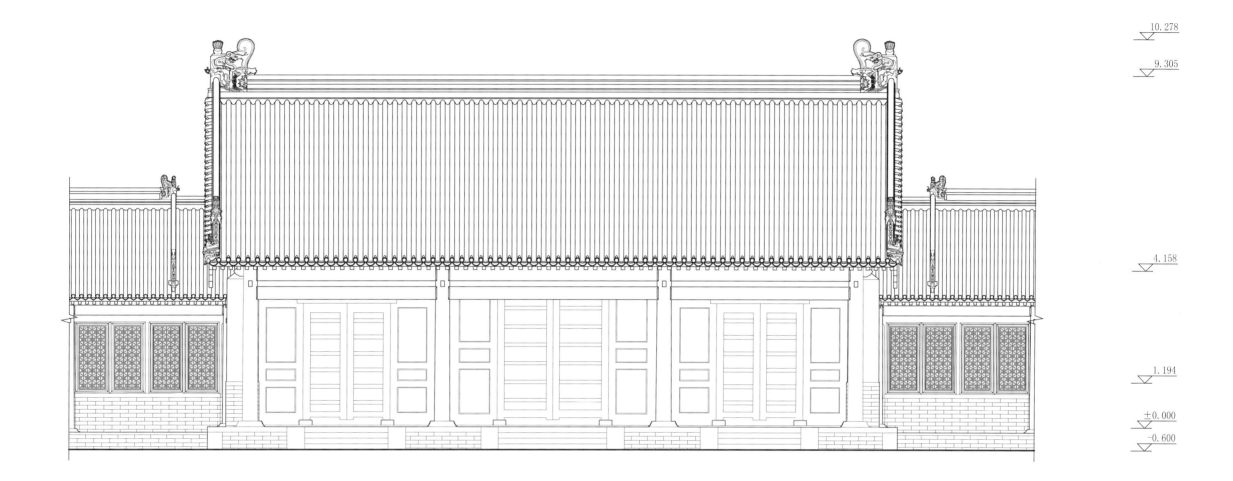

10.278

9.305

4.158

1.194

±0.000

-0.600

1160　　5240　　6080　　5240　　1160

18880

神乐署大门背立面图
Rear elevation of entrance gate of Divine Music Administration

0　　1　　3m

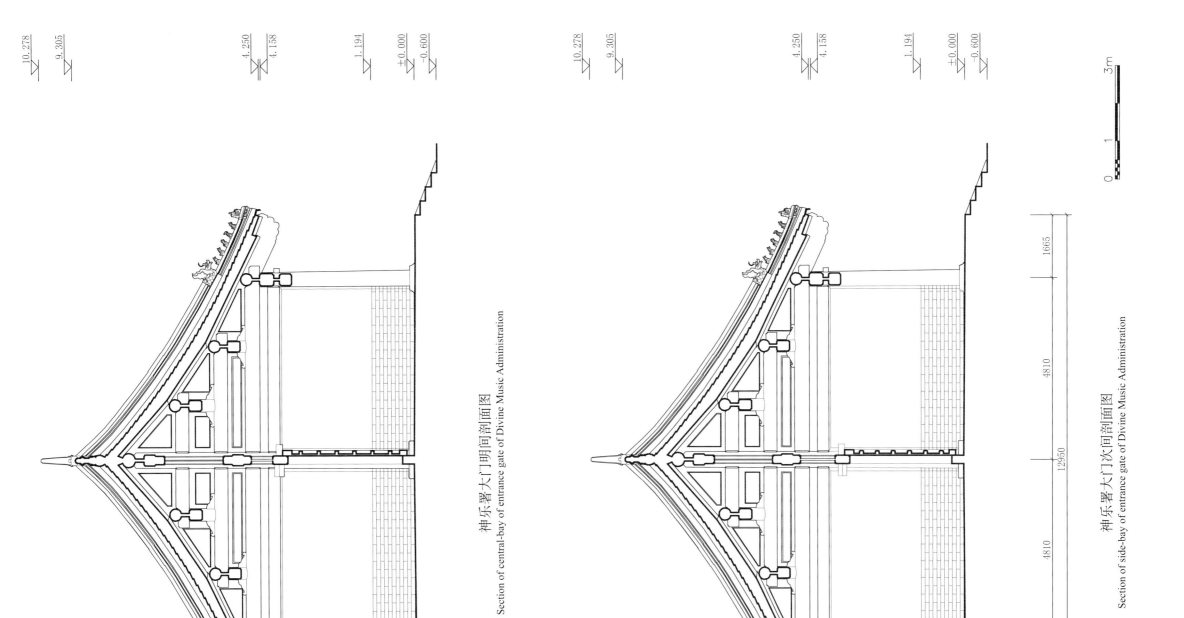

10.278
9.305

4.250
4.158

1.194

±0.000
-0.600

神乐署大门明间剖面图
Section of central-bay of entrance gate of Divine Music Administration

10.278
9.305

4.250
4.158

1.194

±0.000
-0.600

1665

4810

12950

4810

1665

神乐署大门次间剖面图
Section of side-bay of entrance gate of Divine Music Administration

3m

0   1

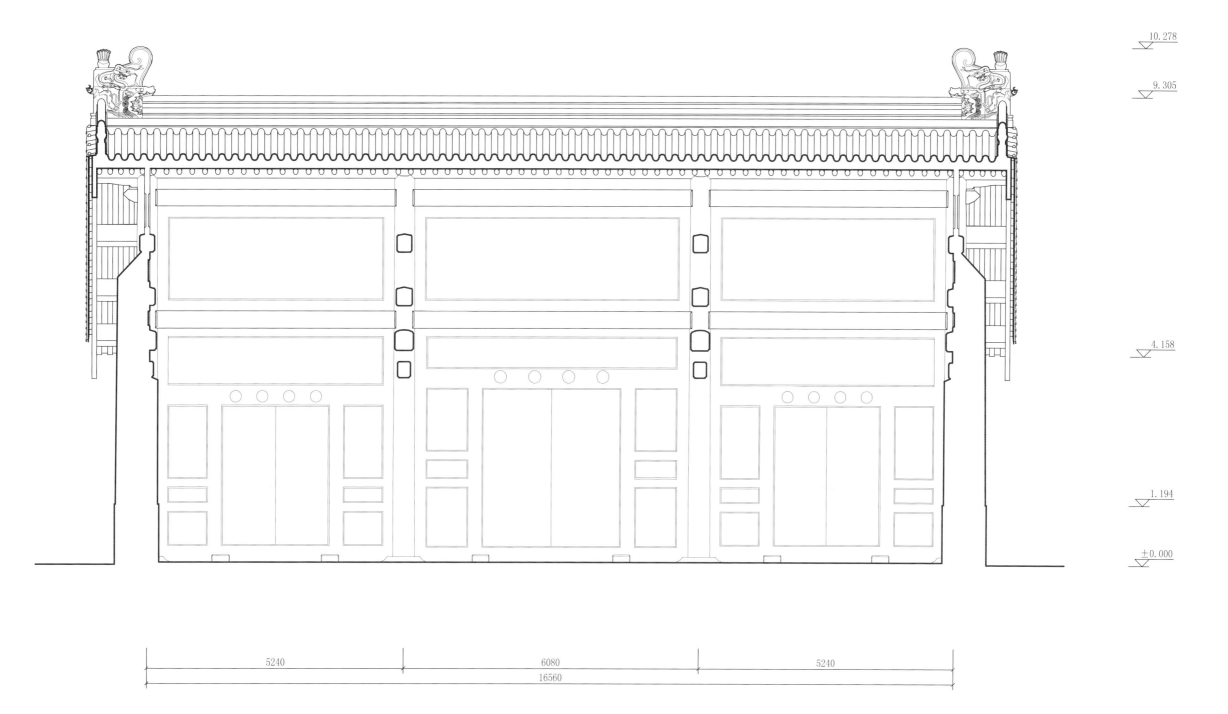

10.278

9.305

4.158

1.194

±0.000

5240　　　　6080　　　　5240

16560

神乐署大门纵剖面图
Longitudinal section of entrance gate of Divine Music Administration

0　　1　　2m

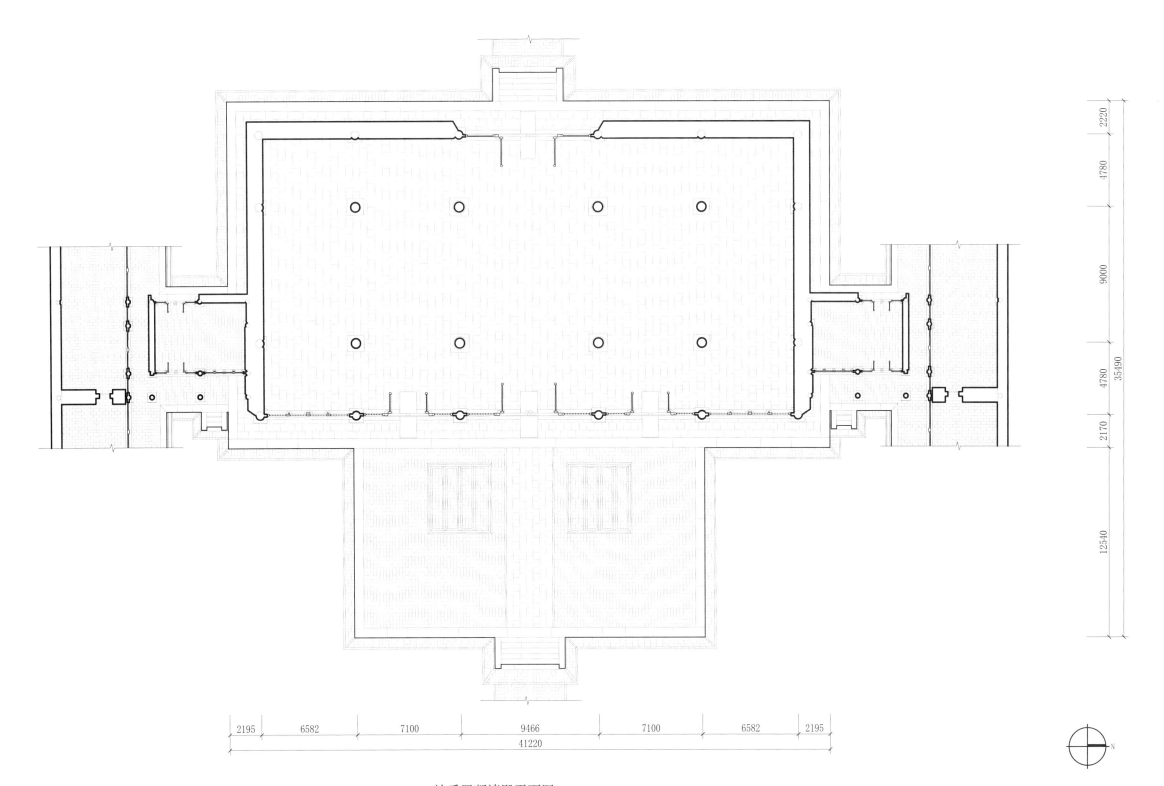

| 2220 | 4780 | 9000 | 4780 | 2170 | 12540 |

35490

| 2195 | 6582 | 7100 | 9466 | 7100 | 6582 | 2195 |

41220

神乐署凝禧殿平面图
Plan of Hall of Accumulated Happiness of Divine Music Administration

0 1 5m

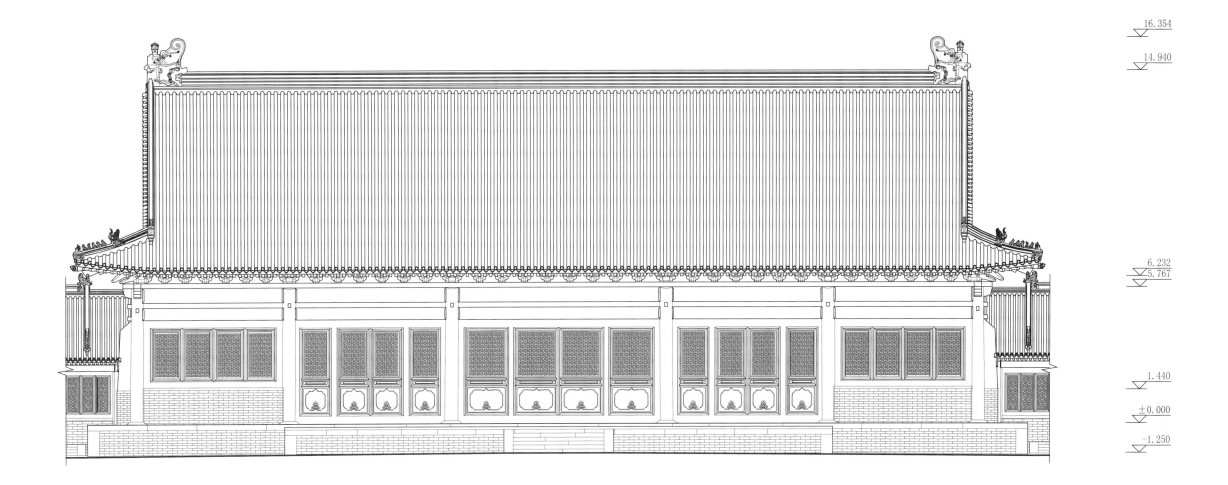

16.354

14.940

6.232
5.767

1.440

±0.000

-1.250

| 2195 | 6582 | 7100 | 9466 | 7100 | 6582 | 2195 |

41220

神乐署凝禧殿正立面图
Front elevation of Hall of Accumulated Happiness of Divine Music Administration

0 1   3m

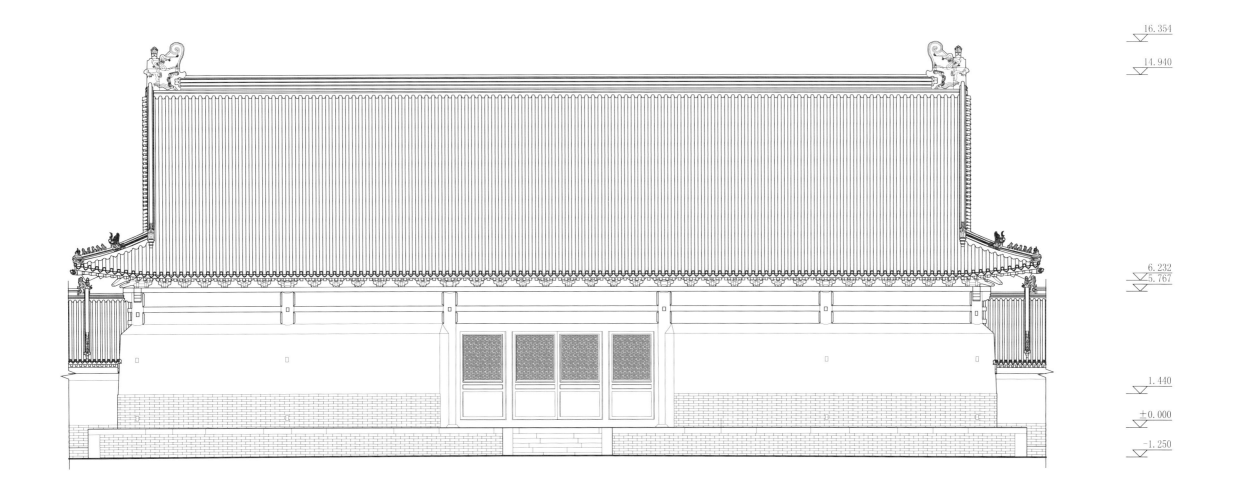

16.354

14.940

6.232
5.767

1.440

±0.000

-1.250

| 2195 | 6582 | 7100 | 9466 | 7100 | 6582 | 2195 |

41220

神乐署凝禧殿背立面图
Rear elevation of Hall of Accumulated Happiness of Divine Music Administration

0 1 3m

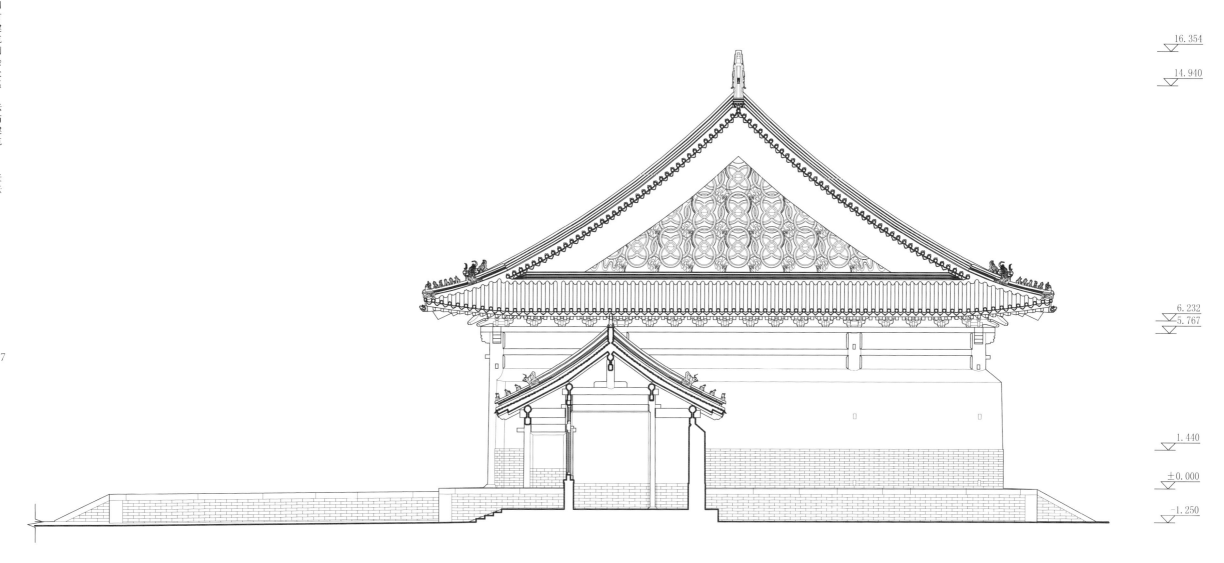

16.354

14.940

6.232
5.767

1.440

±0.000

-1.250

| 12540 | 2230 | 4755 | 8930 | 4755 | 2280 |

35490

神乐署凝禧殿侧立面图
Side elevation of Hall of Accumulated Happiness of Divine Music Administration

0 1 3m

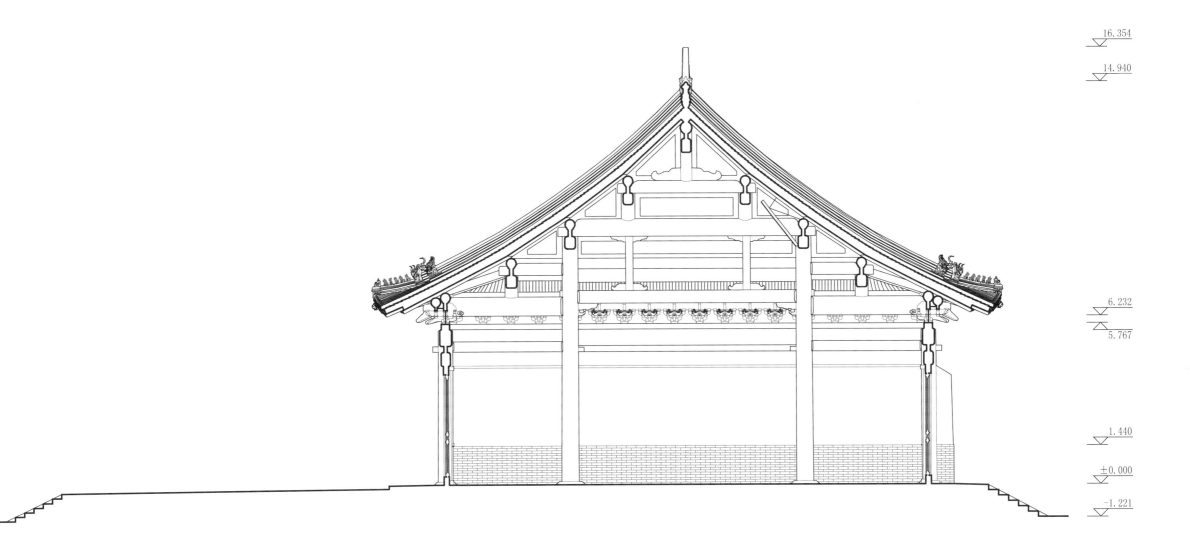

16.354

14.940

6.232

5.767

1.440

±0.000

-1.221

12540　　2230　　4755　　8930　　4755　　2280

35490

神乐署凝禧殿明间剖面图
Section of central-bay of Hall of Accumulated Happiness of Divine Music Administration

0　1　　3m

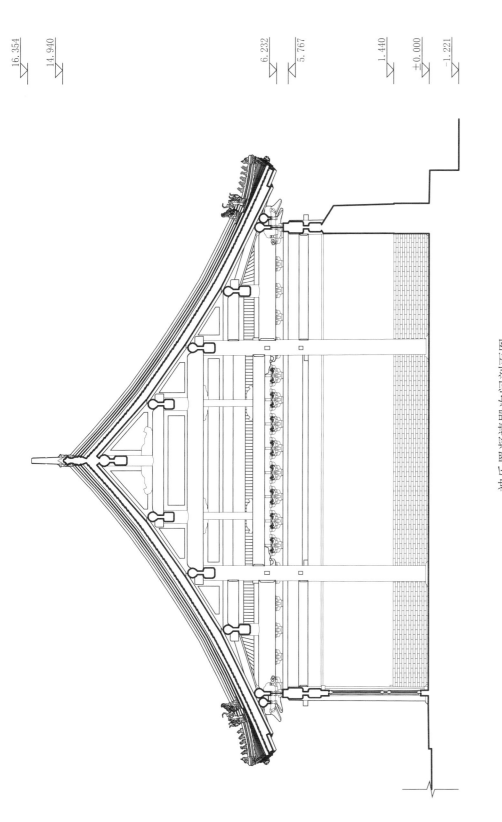

神乐署凝禧殿次间剖面图
Section of side-bay of Hall of Accumulated Happiness of Divine Music Administration

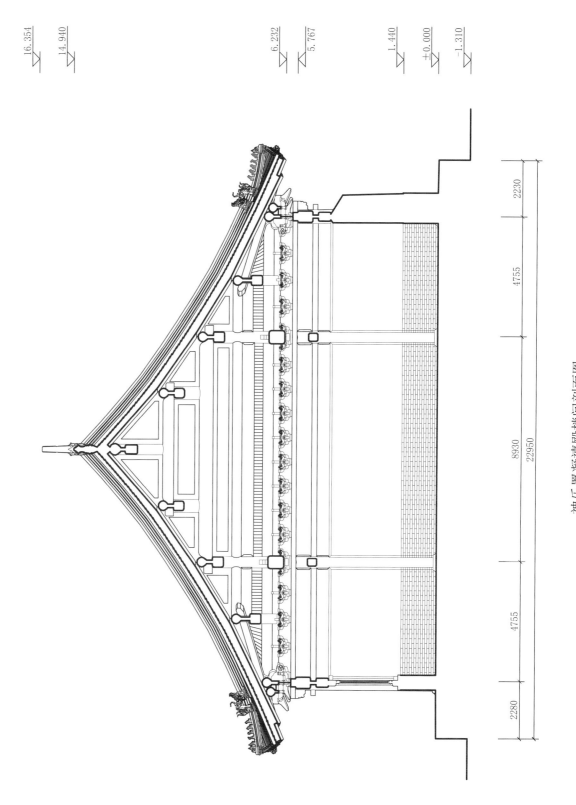

神乐署凝禧殿稍间剖面图
Section of second-to-last-bay of Hall of Accumulated Happiness of Divine Music Administration

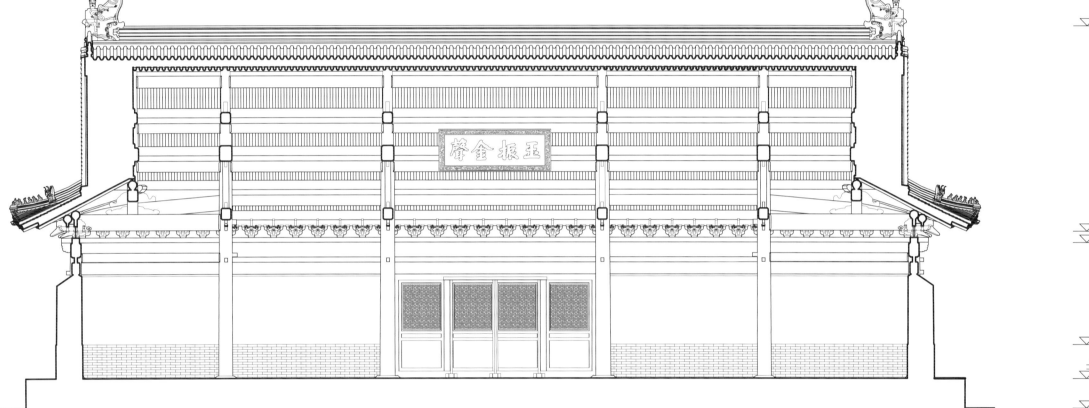

16.354

14.940

6.232
5.767

1.440

±0.000

-1.250

| 2195 | 6582 | 7100 | 9466 | 7100 | 6582 | 2195 |

41220

神乐署凝禧殿纵剖面图
Longitudinal section of Hall of Accumulated Happiness of Divine Music Administration

0 1 3m

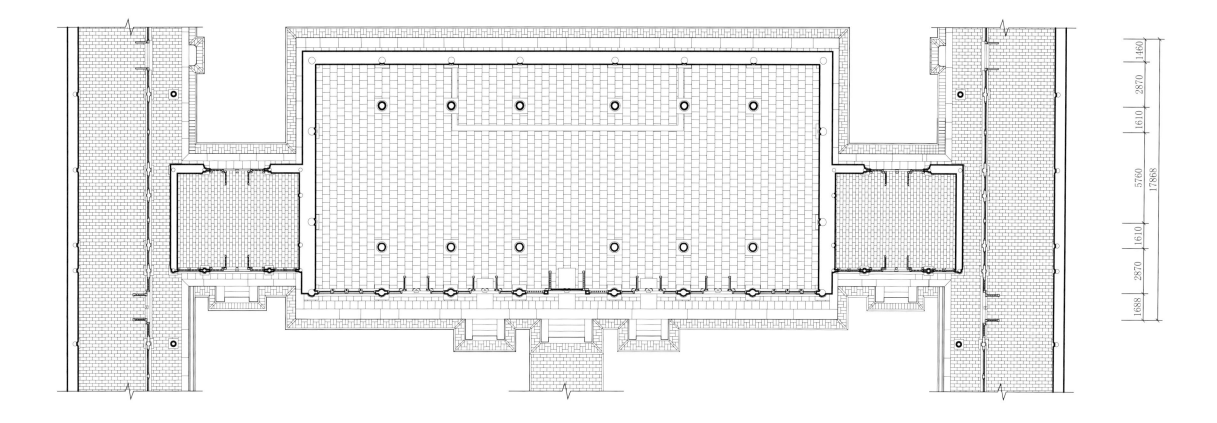

神乐署显佑殿平面图
Plan of Xianyou Hall of Divine Music Administration

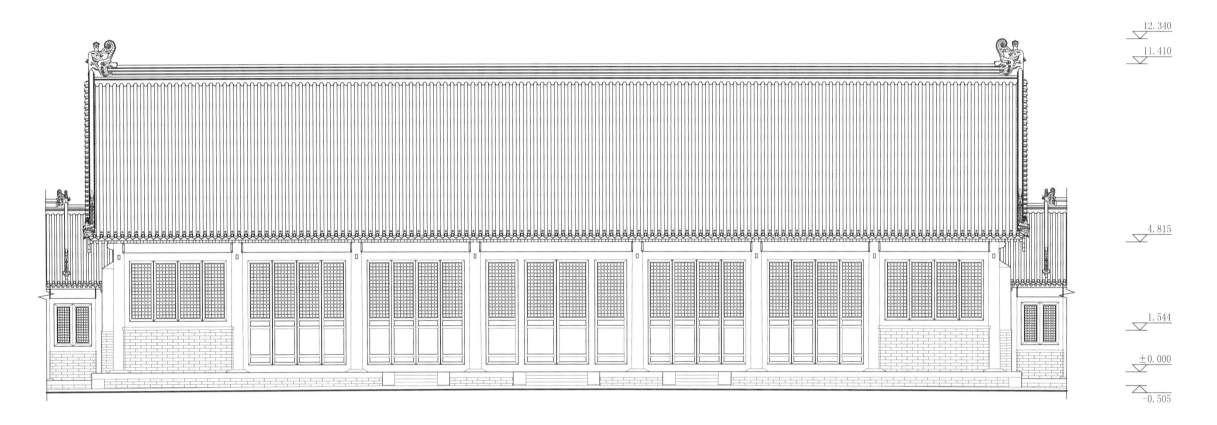

12.340
11.410
4.815
1.544
±0.000
-0.505

1030	4560	4560	4560	6220	4560	4560	4560	1030

35640

神乐署显佑殿正立面图
Front elevation of Xianyou Hall of Divine Music Administration

0 1 3m

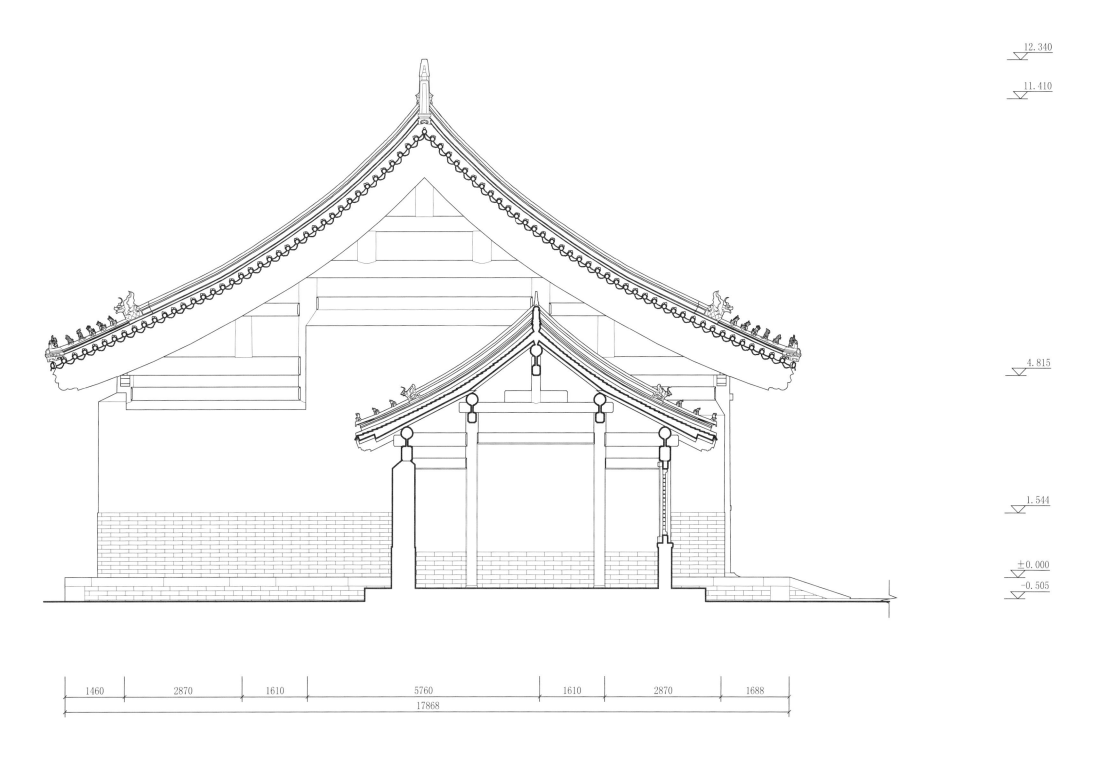

12.340

11.410

4.815

1.544

±0.000

-0.505

| 1460 | 2870 | 1610 | 5760 | 1610 | 2870 | 1688 |

17868

神乐署显佑殿侧立面图
Side elevation of Xianyou Hall of Divine Music Administration

0   1   3m

神乐署显佑殿明间剖面图
Section of central-bay of Xianyou Hall of Divine Music Administration

神乐署显佑殿梢间剖面图
Section of second-to-last-bay of Xianyou Hall of Divine Music Administration

0 1 3m

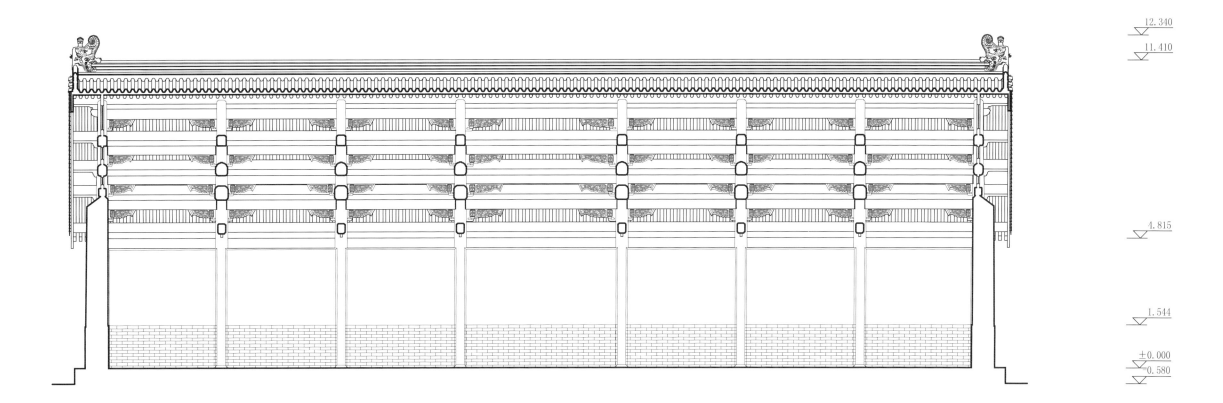

12.340

11.410

4.815

1.544

±0.000

-0.580

| 1030 | 4560 | 4560 | 4560 | 6220 | 4560 | 4560 | 4560 | 1030 |

35640

神乐署显佑殿纵剖面图
Longitudinal section of Xianyou Hall of Divine Music Administration

0   1   3m

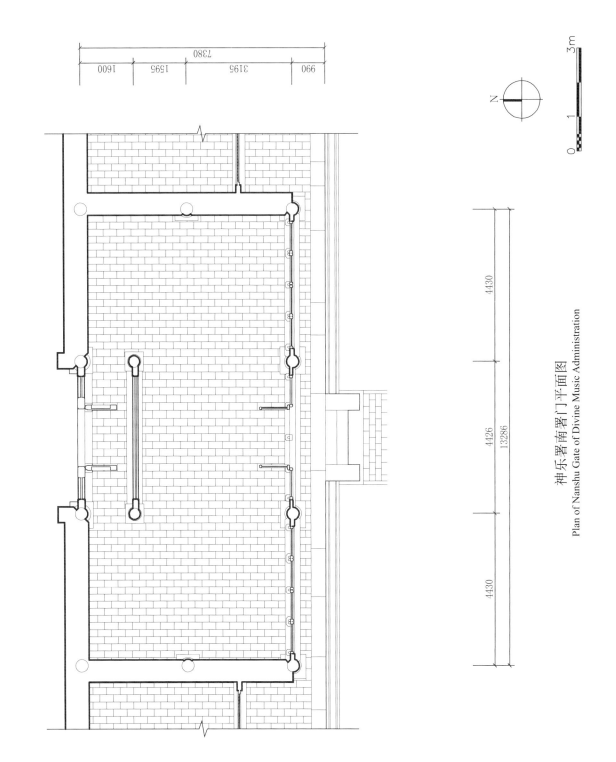

神乐署南署门署门正面图
Front elevation of Nanshu Gate of Divine Music Administration

神乐署南署门平面图
Plan of Nanshu Gate of Divine Music Administration

8.302
7.700
6.550
3.863
0.967
±0.000
0.447

1600
1595
3195
990
7380

4430
4426
13286
4430

N

0 1 3m

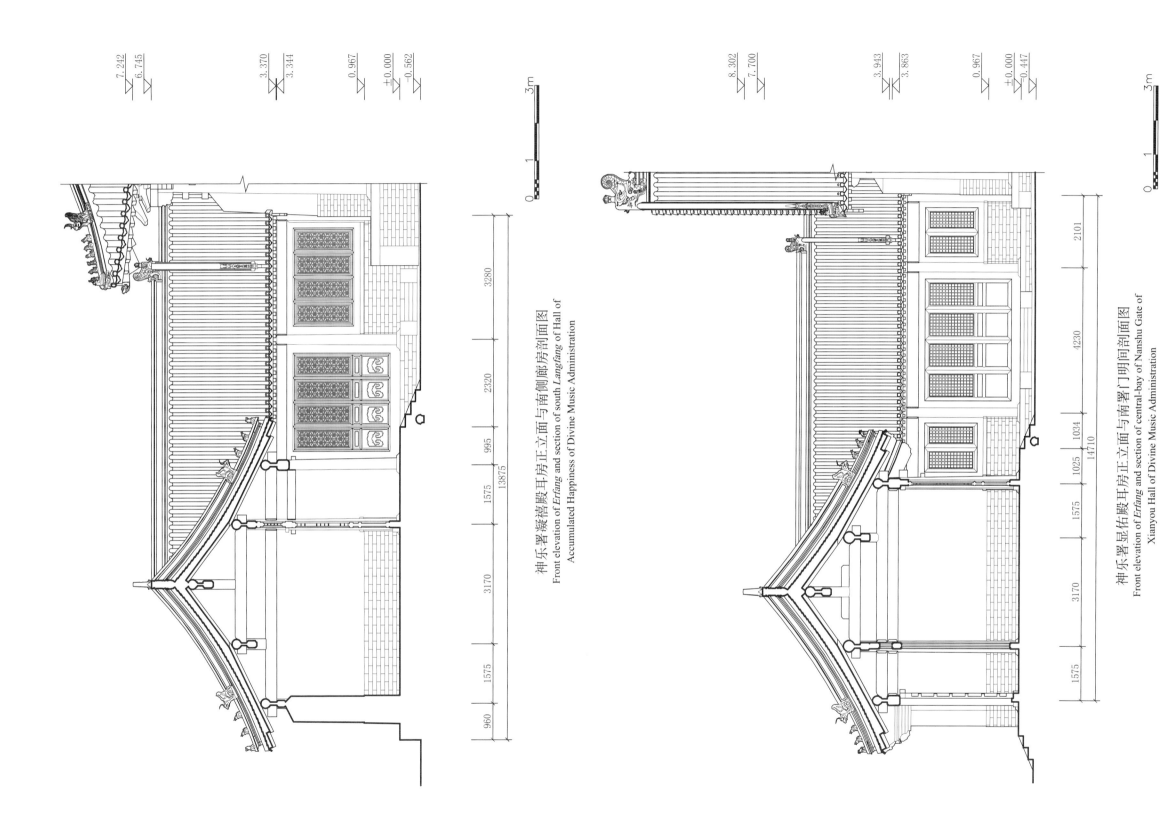

神乐署凝禧殿耳房正立面与南署侧廊房剖面图
Front elevation of *Erfang* and section of south *Langfang* of Hall of
Accumulated Happiness of Divine Music Administration

神乐署显佑殿耳房正立面与南署门明间剖面图
Front elevation of *Erfang* and section of central-bay of Nanshu Gate of
Xianyou Hall of Divine Music Administration

1 院门　Entrance Gate
2 神库　Divine Storeroom
3 西神厨　West Divine Storeroom
4 东神厨　East Divine Storeroom
5 井庭　*Jingting*

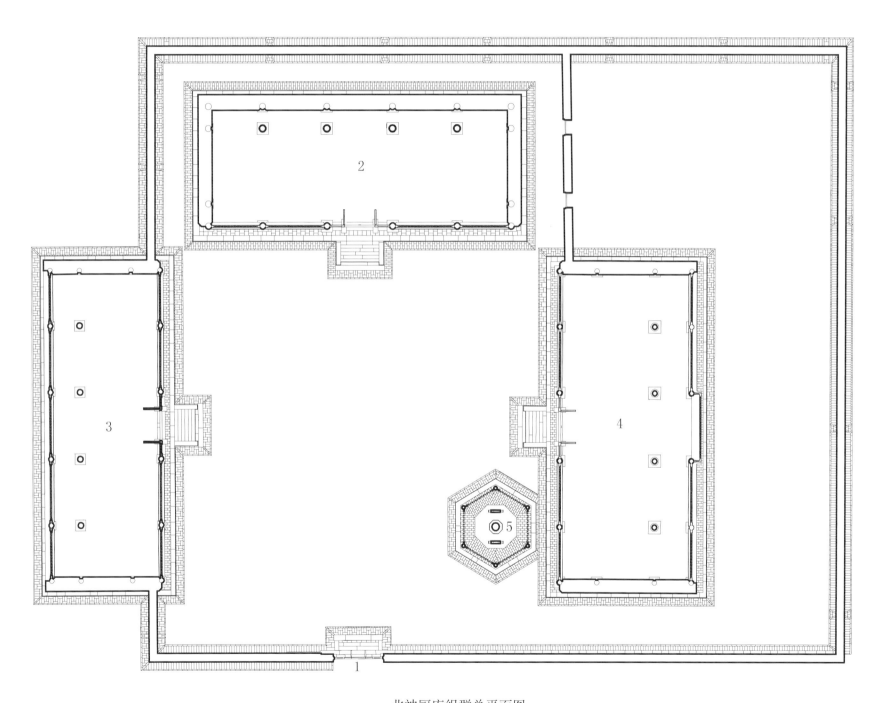

北神厨库组群总平面图
Site plan of Complex of North Divine Kitchen and Divine Storeroom

N

0 1　　5　　10m

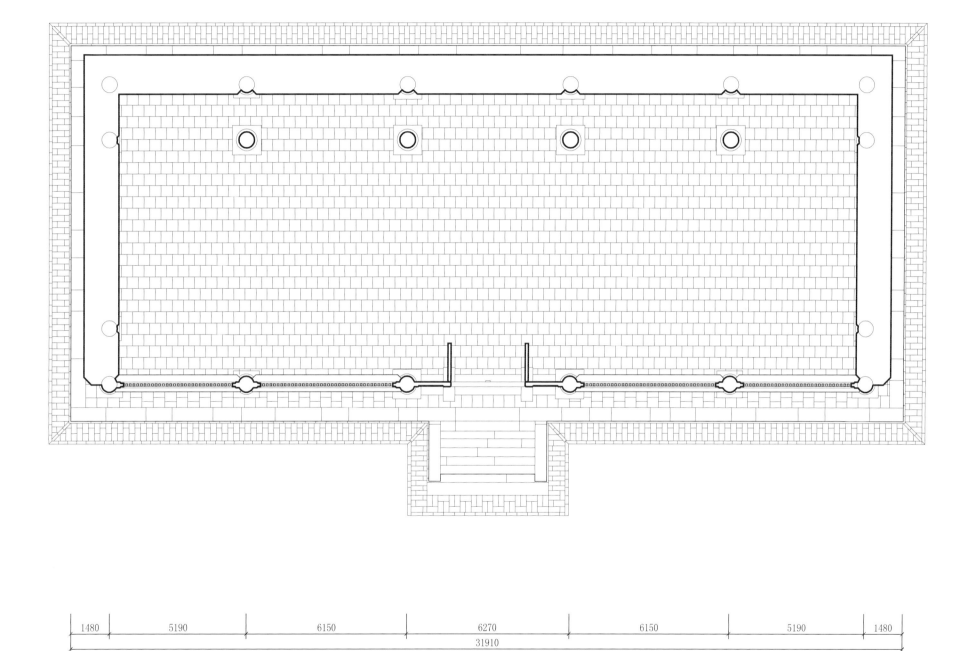

| 1480 | 5190 | 6150 | 6270 | 6150 | 5190 | 1480 |

31910

北神厨库神库平面图
Plan of Divine Storeroom of North Divine Kitchen and Divine Storeroom

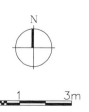

0   1   3m

10.890

9.895

4.220

190

±0.000

−1.175

| 1480 | 5190 | 6150 | 6270 | 6150 | 5190 | 1480 |

31910

北神厨库神库正立面图

Front elevation of Divine Storeroom of North Divine Kitchen and Divine Storeroom

0 1   3m

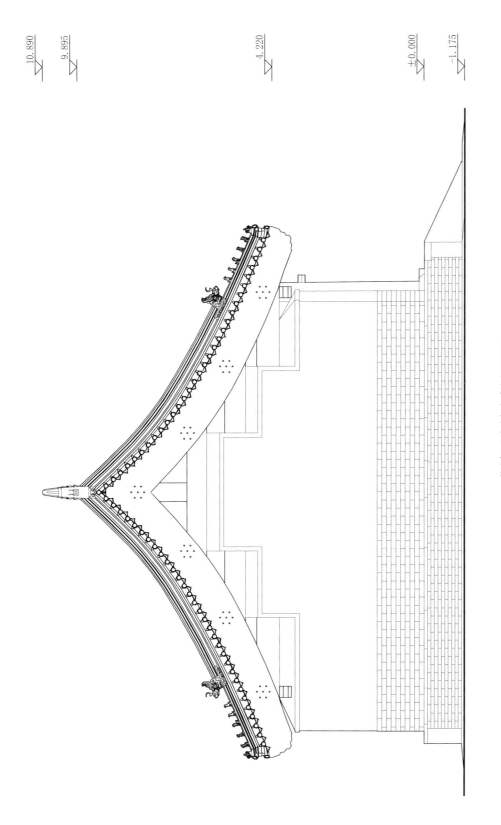

北神厨库神库侧立面图
Side elevation of Divine Storeroom of North Divine Kitchen and Divine Storeroom

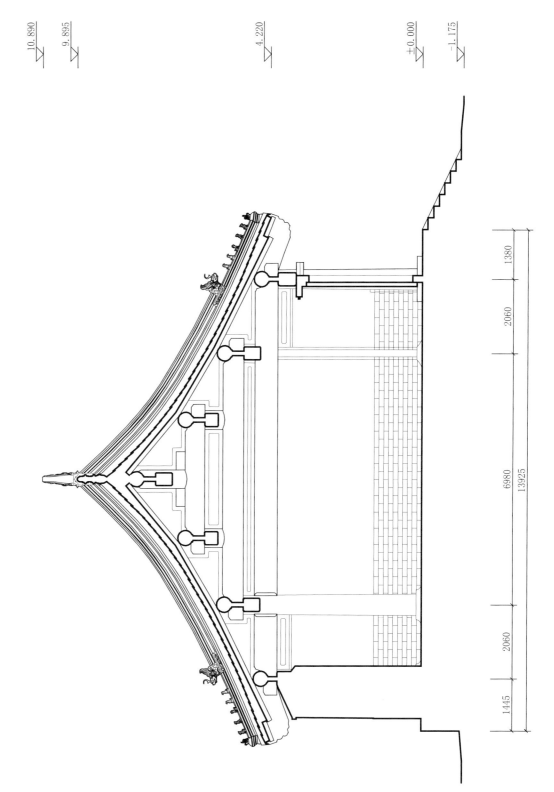

北神厨库神库明间剖面图
Section of central-bay of Divine Storeroom of North Divine Kitchen and Divine Storeroom

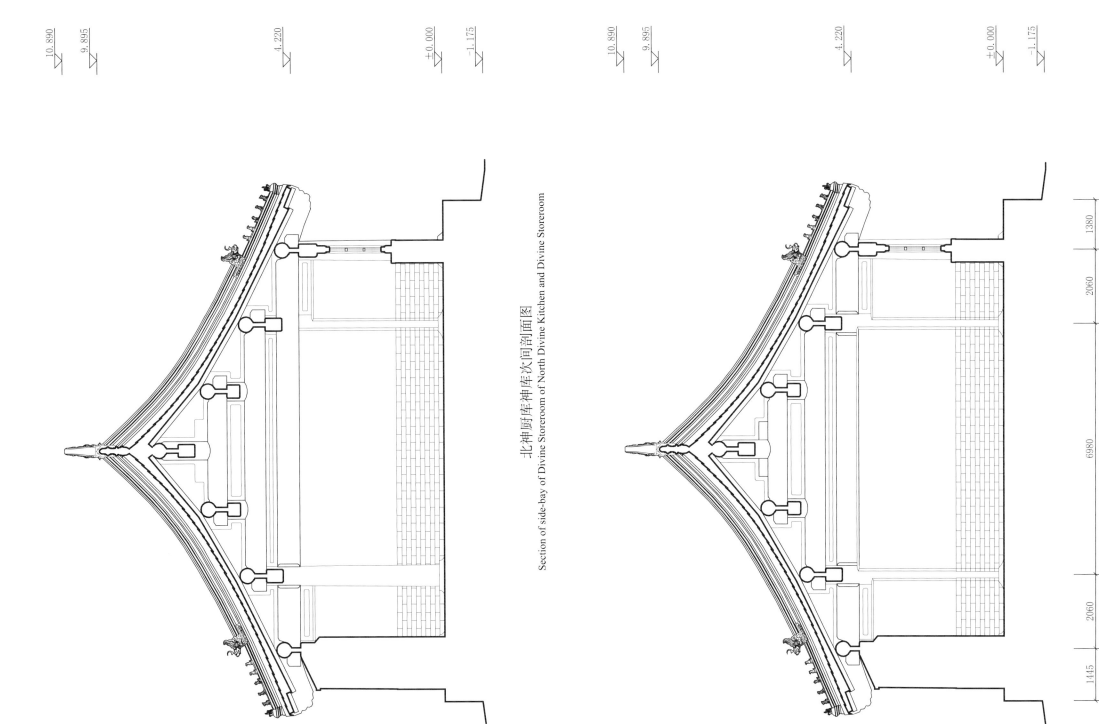

10.890

9.895

4.220

±0.000

-1.175

北神厨库神库次间剖面图
Section of side-bay of Divine Storeroom of North Divine Kitchen and Divine Storeroom

10.890

9.895

4.220

±0.000

-1.175

1380

2060

6980

13925

2060

1445

北神厨库神库梢间剖面图
Section of second-to-last-bay of Divine Storeroom of North Divine Kitchen and Divine Storeroom

0　　1　　3m

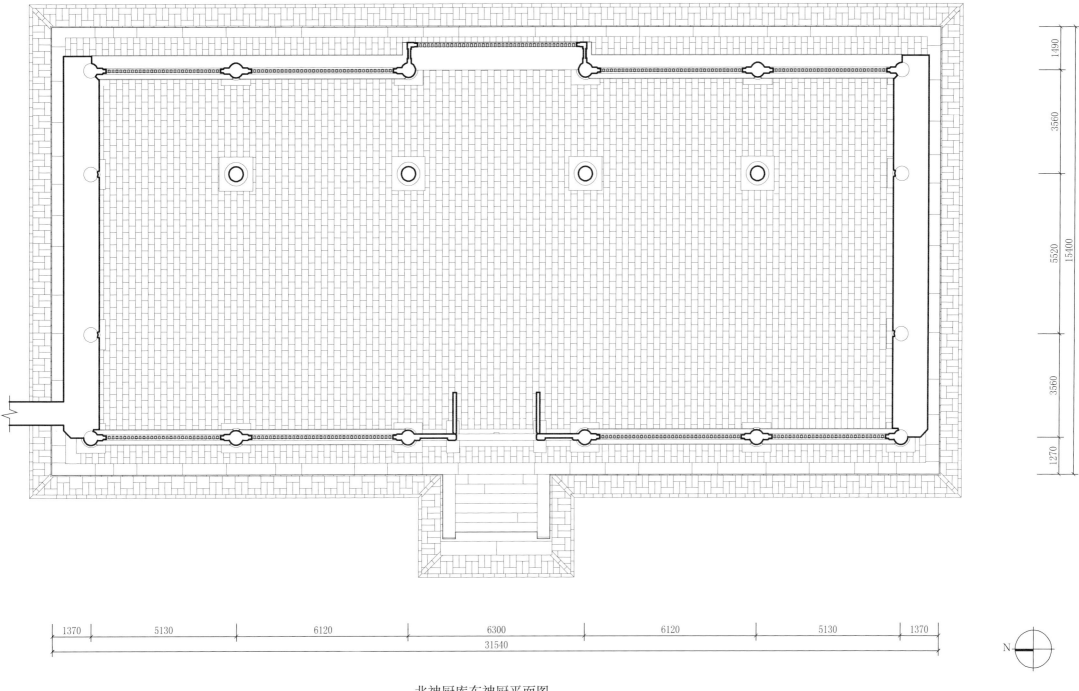

1490
3560
5520
15400
3560
1270

1370 5130 6120 6300 6120 5130 1370
31540

北神厨库东神厨平面图
Plan of east Divine kitchen of North Divine Kitchen and Divine Storeroom

N

0 1 3m

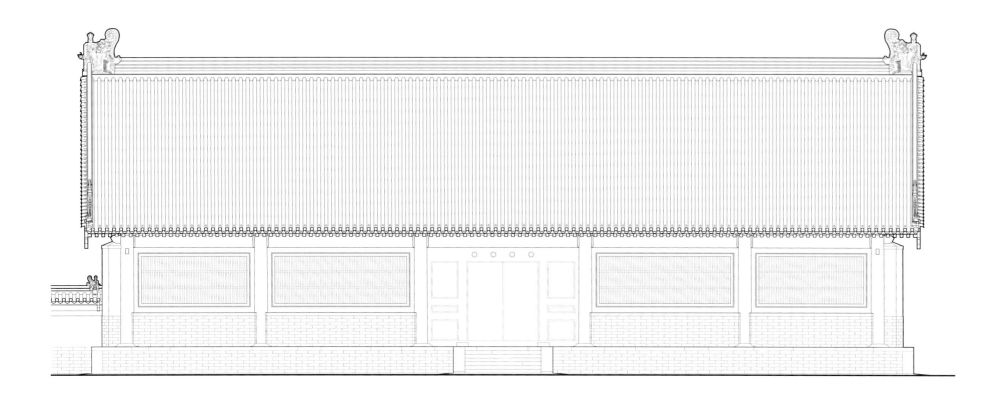

11.790

10.715

4.070

±0.000

-1.045

| 1370 | 5130 | 6120 | 6300 | 6120 | 5130 | 1370 |

31540

北神厨库东神厨正立面图

Front elevation of east Divine kitchen of North Divine Kitchen and Divine Storeroom

0 1 3m

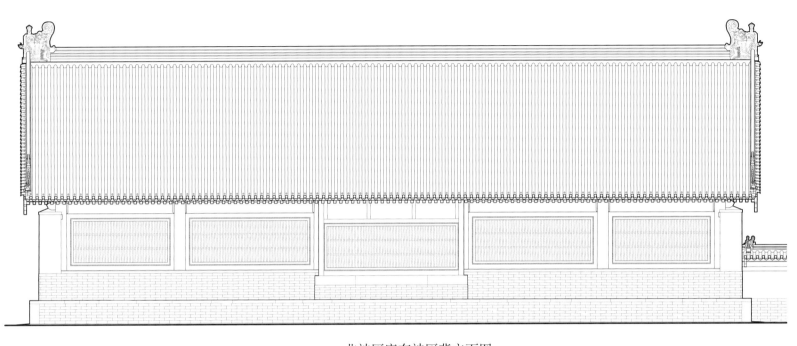

11.790

10.715

4.070

2.275

±0.000

-1.045

北神厨库东神厨背立面图

Rear elevation of east Divine kitchen of North Divine Kitchen and Divine Storeroom

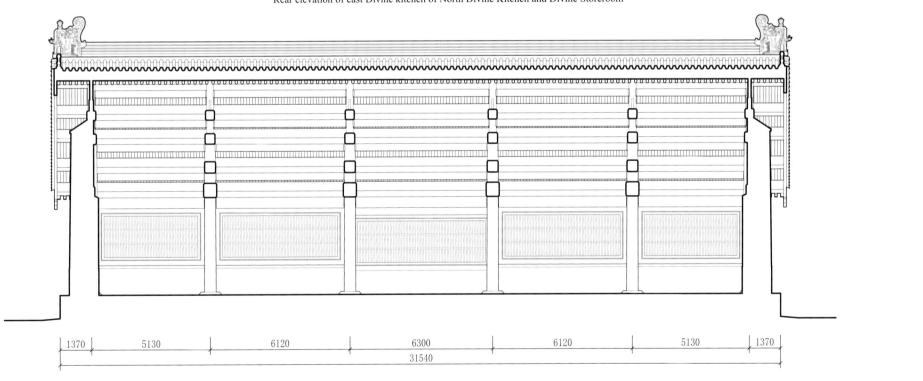

11.790

10.715

4.070

±0.000

-1.045

| 1370 | 5130 | 6120 | 6300 | 6120 | 5130 | 1370 |

31540

北神厨库东神厨纵剖面图

Longitudinal section of east Divine kitchen of North Divine Kitchen and Divine Storeroom

0 1 3m

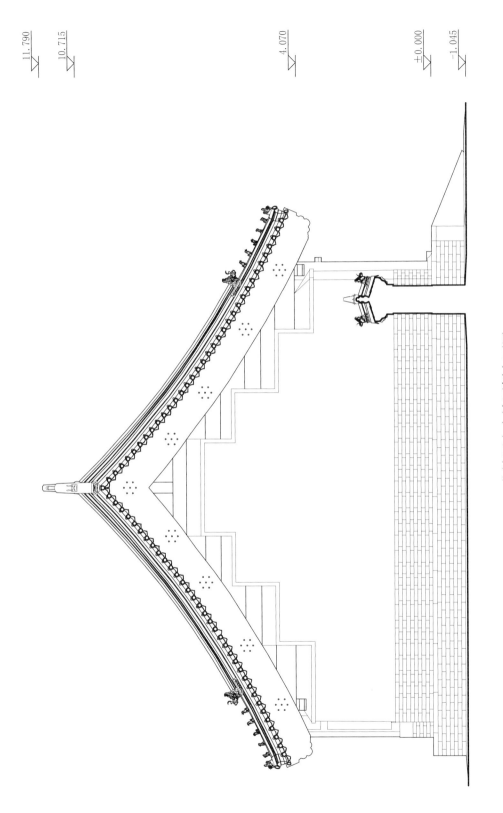

北神厨库东神厨侧立面图
Side elevation of east Divine kitchen of North Divine Kitchen and Divine Storeroom

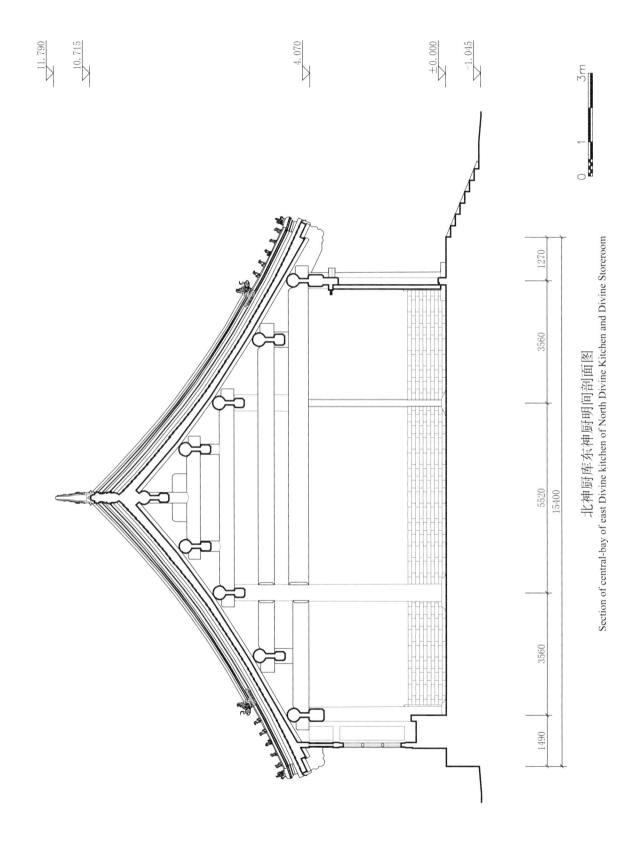

北神厨库东神厨明间剖面图
Section of central-bay of east Divine kitchen of North Divine Kitchen and Divine Storeroom

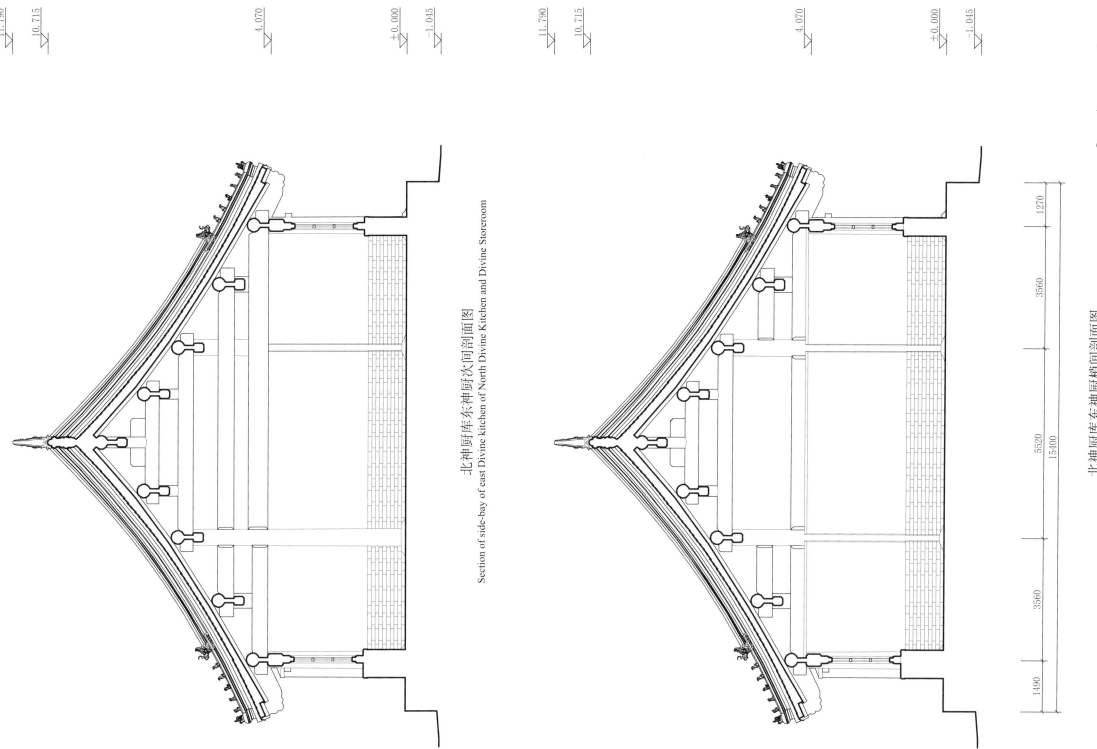

北神厨库东神厨次间剖面图
Section of side-bay of east Divine kitchen of North Divine Kitchen and Divine Storeroom

北神厨库东神厨梢间剖面图
Section of second-to-last-bay of east Divine kitchen of North Divine Kitchen and Divine Storeroom

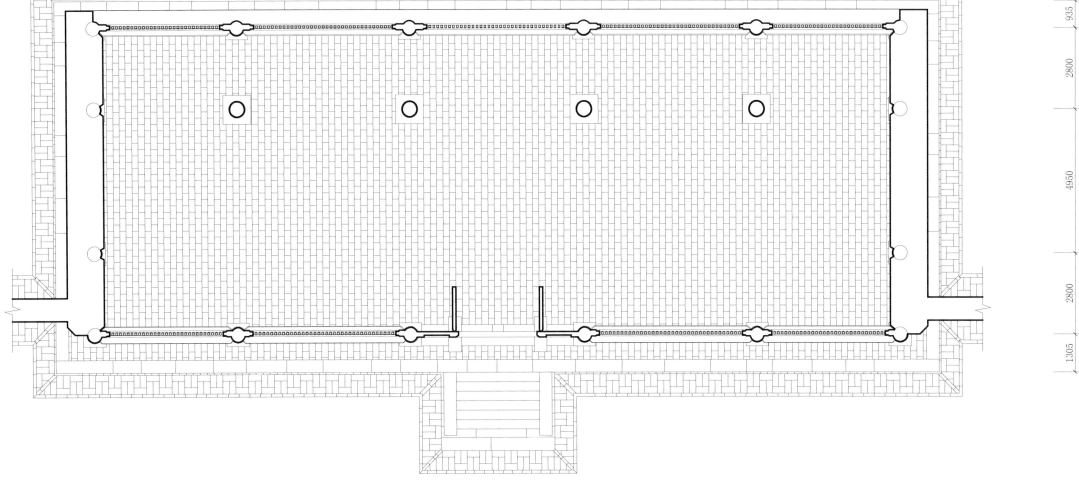

935 2800 4950 12790 2800 1305

1375　5120　6140　6200　6140　5120　1375
31470

北神厨库西神厨平面图
Plan of west Divine kitchen of North Divine Kitchen and Divine Storeroom

0　1　3m

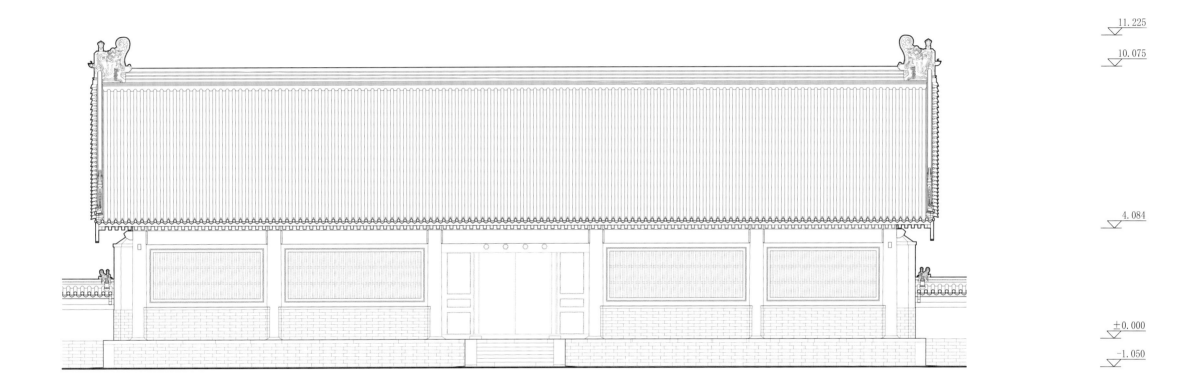

11.225

10.075

4.084

±0.000

-1.050

| 1375 | 5120 | 6140 | 6200 | 6140 | 5120 | 1375 |

31470

北神厨库西神厨正立面图
Front elevation of west Divine kitchen of North Divine Kitchen and Divine Storeroom

0 1 3m

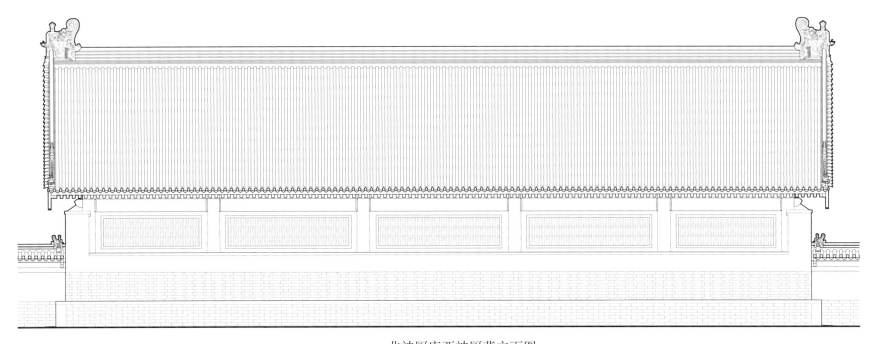

11.790

10.715

4.070

2.275

±0.000

-1.045

北神厨库西神厨背立面图
Rear elevation of west Divine kitchen of North Divine Kitchen and Divine Storeroom

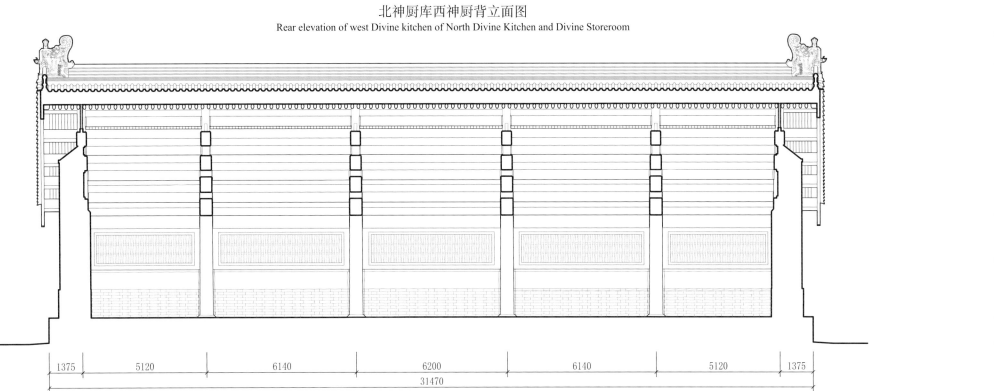

11.790

10.715

4.070

2.275

±0.000

-1.045

| 1375 | 5120 | 6140 | 6200 | 6140 | 5120 | 1375 |

31470

北神厨库西神厨纵剖面图
Longitudinal section of west Divine kitchen of North Divine Kitchen and Divine Storeroom

0 1 3m

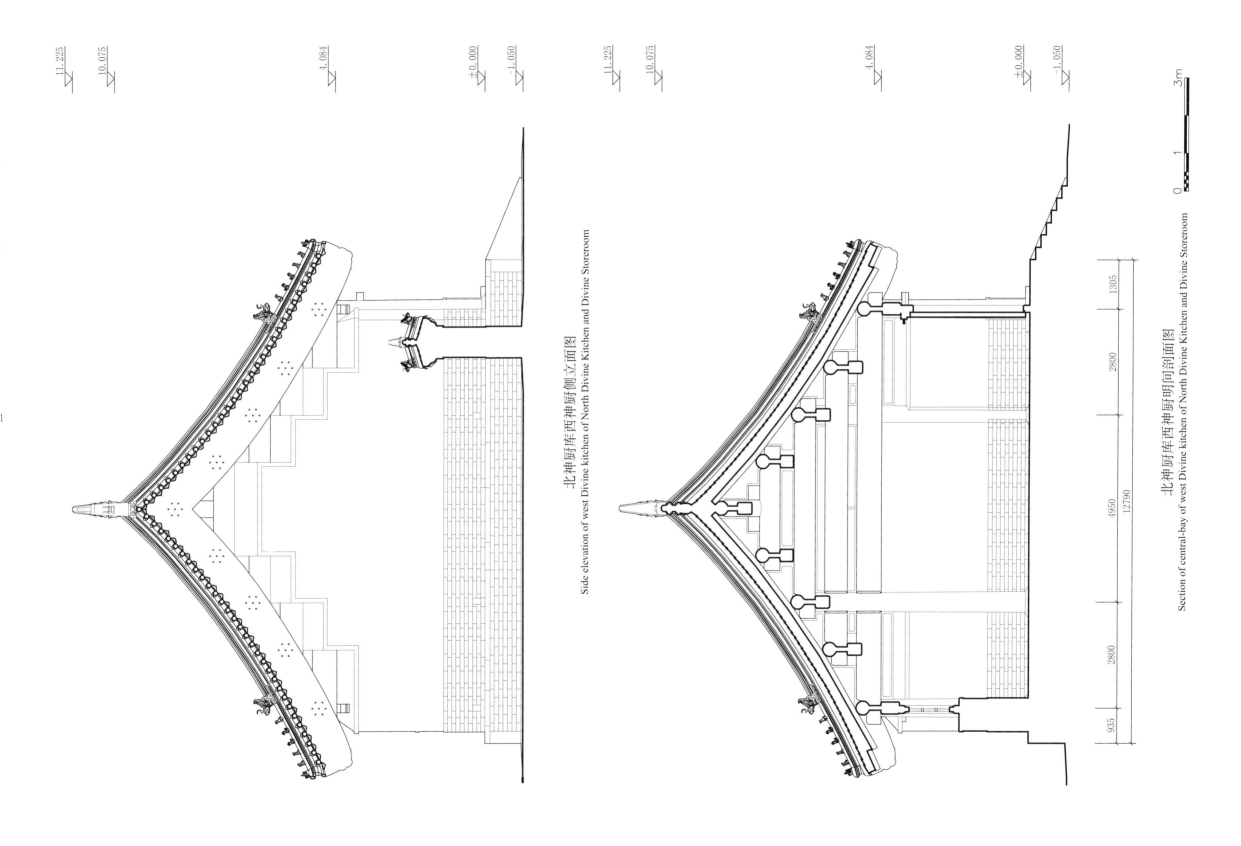

11.225
10.075

4.084

±0.000
-1.050

北神厨库西神厨侧立面图
Side elevation of west Divine kitchen of North Divine Kitchen and Divine Storeroom

11.225
10.075

4.084

±0.000
-1.050

1305

2800

4950   12790

2800

935

北神厨库西神厨明间剖面图
Section of central-bay of west Divine kitchen of North Divine Kitchen and Divine Storeroom

0   1   3m

北神厨库西神厨次间剖面图
Section of side-bay of west Divine kitchen of North Divine Kitchen and Divine Storeroom

北神厨库西神厨梢间剖面图
Section of second-to-last-bay of west Divine kitchen of North Divine Kitchen and Divine Storeroom

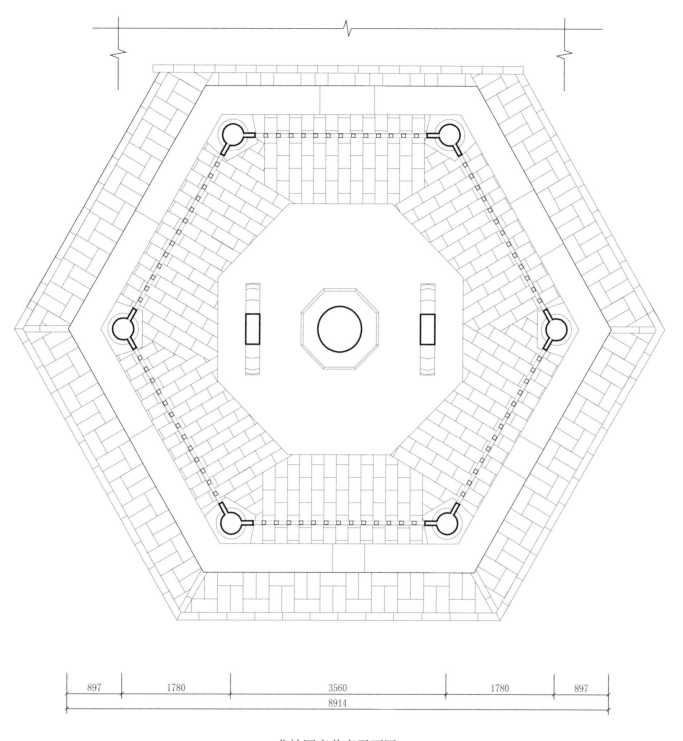

897 | 1780 | 3560 | 1780 | 897

8914

777

3083

7720

3083

777

北神厨库井亭平面图
Plan of *Jingting* of North Divine Kitchen and Divine Storeroom

N

0   1   2m

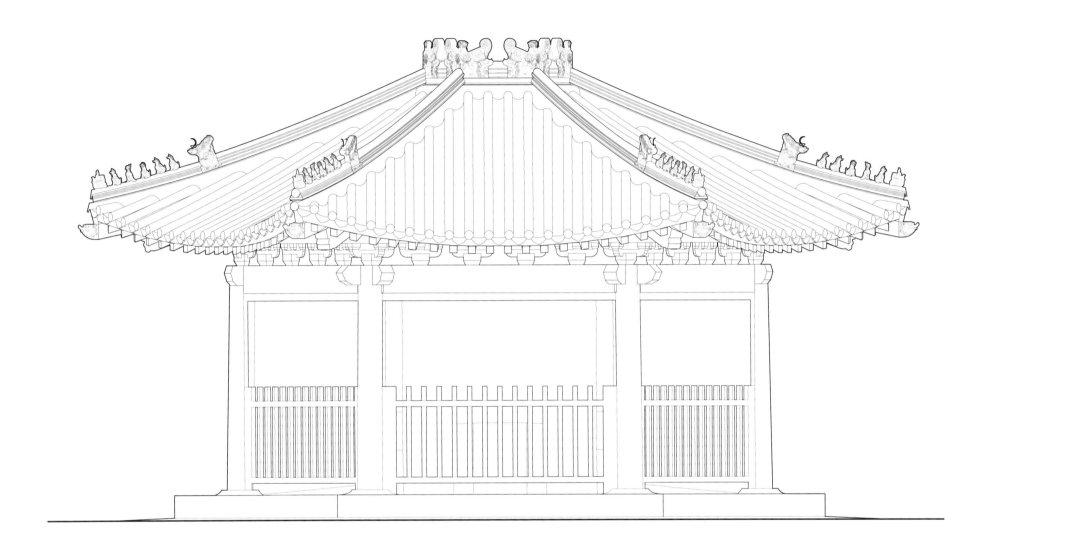

6.030

5.730

3.035

±0.000

-0.335

897   1780   3560   1780   897

8914

北神厨库井亭正立面图
Front elevation of *Jingting* of North Divine Kitchen and Divine Storeroom

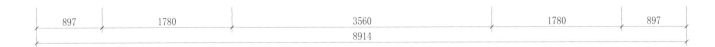

0   1   2m

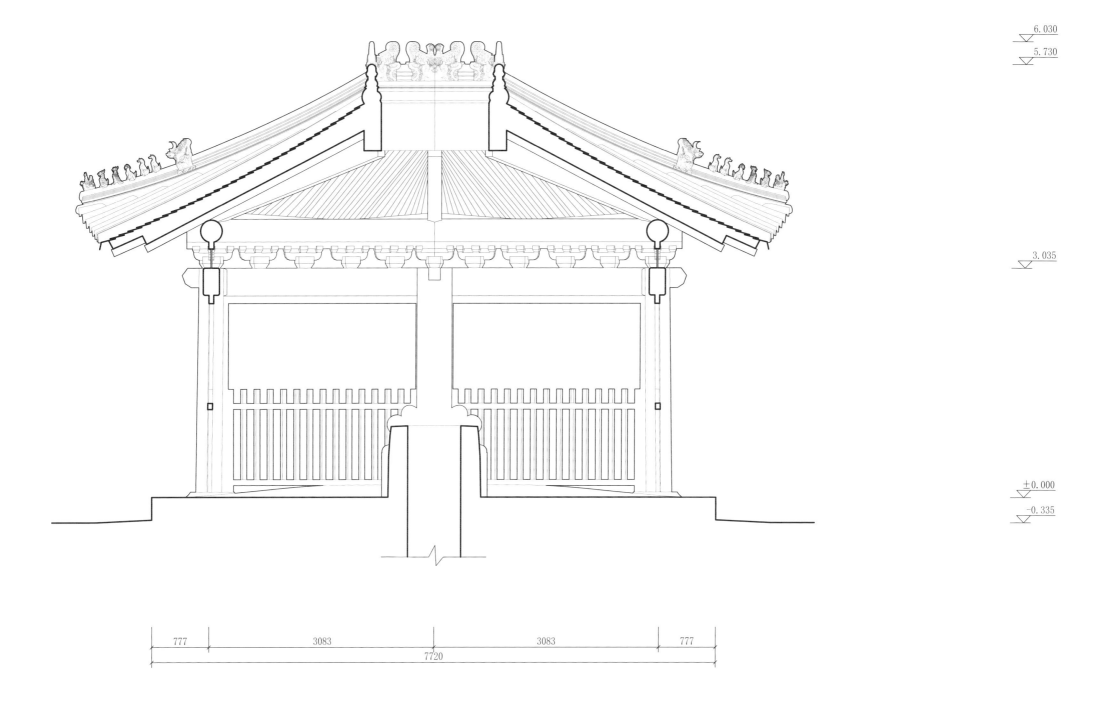

6.030

5.730

3.035

±0.000

-0.335

| 777 | 3083 | 3083 | 777 |
| 7720 | | | |

北神厨库井亭横剖面图
Cross-section of *Jingting* of North Divine Kitchen and Divine Storeroom

0　　　1　　　2m

1　北宰牲亭　North Pavilion of Immolation
2　门殿　*Mendian*
3　井亭　*Jingting*

1

2

3

北宰牲亭组群院落平面图
Plan of courtyard of Complex of North Pavilion of Immolation

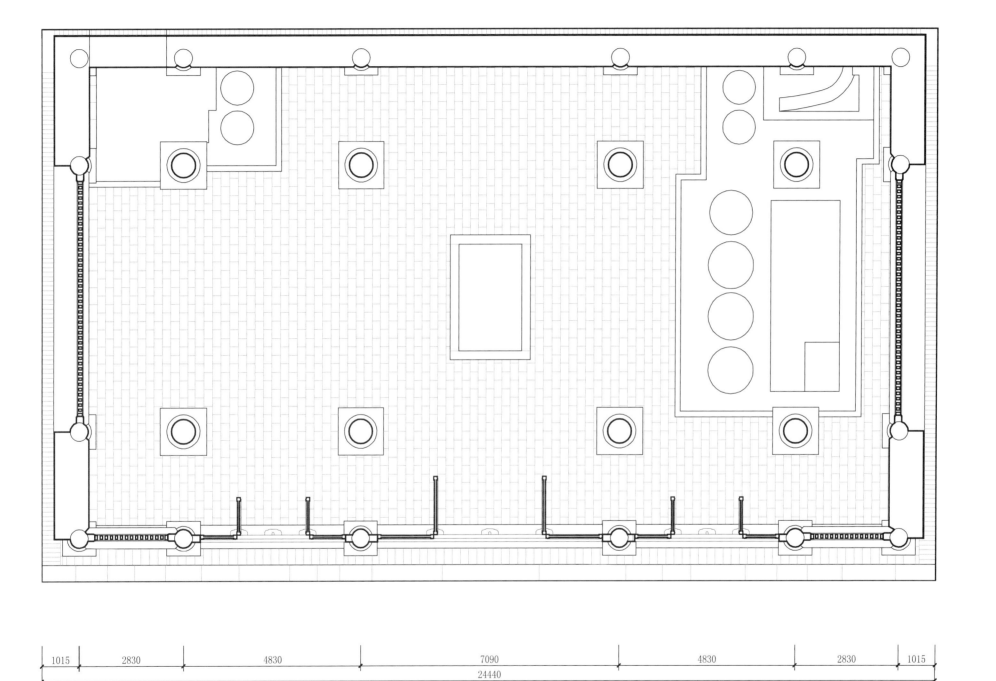

780
2830
7045
14620
2830
1135

1015　2830　4830　7090　4830　2830　1015
24440

北宰牲亭平面图
Plan of North Pavilion of Immolation

N

0　1　3m

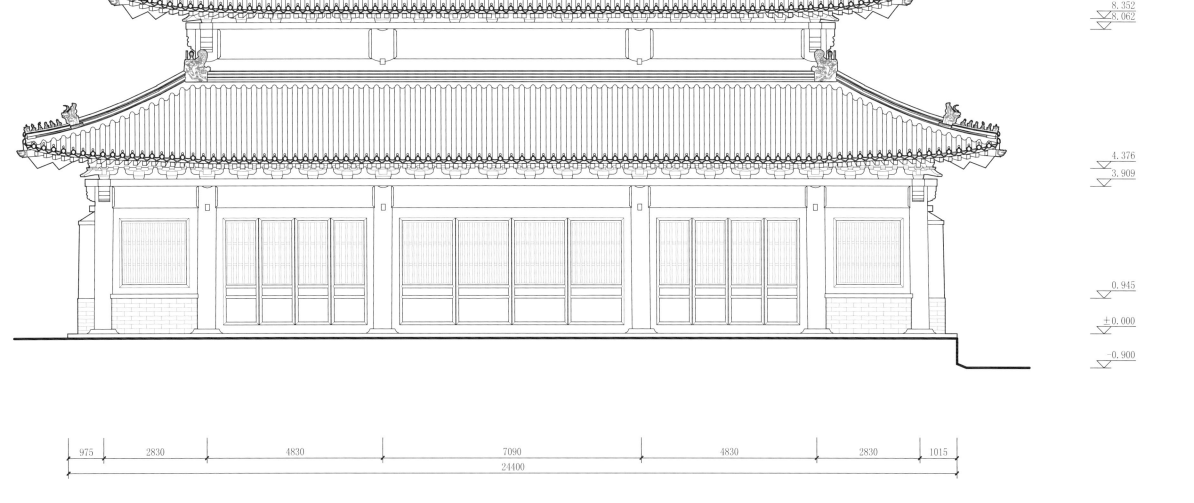

13.449

12.719

8.352
8.062

4.376
3.909

0.945

±0.000

-0.900

| 975 | 2830 | 4830 | 7090 | 4830 | 2830 | 1015 |
| 24400 | | | | | | |

北宰牲亭正立面图
Front elevation of North Pavilion of Immolation

0   1   3m

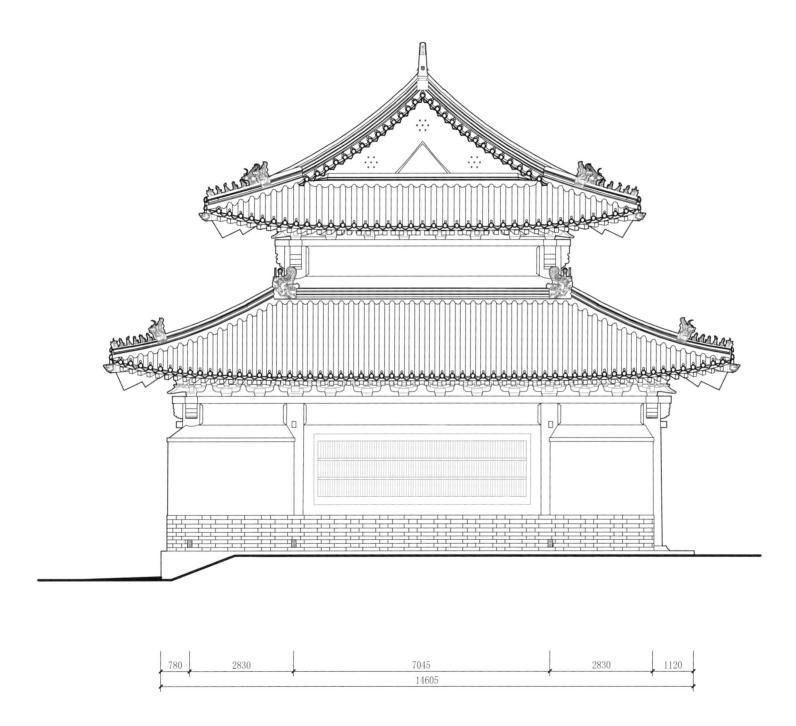

13.449
12.719
8.352
8.062
4.376
3.909
0.945
±0.000
-0.753

| 780 | 2830 | 7045 | 2830 | 1120 |

14605

北宰牲亭侧立面图
Side elevation of North Pavilion of Immolation

0   1   3m

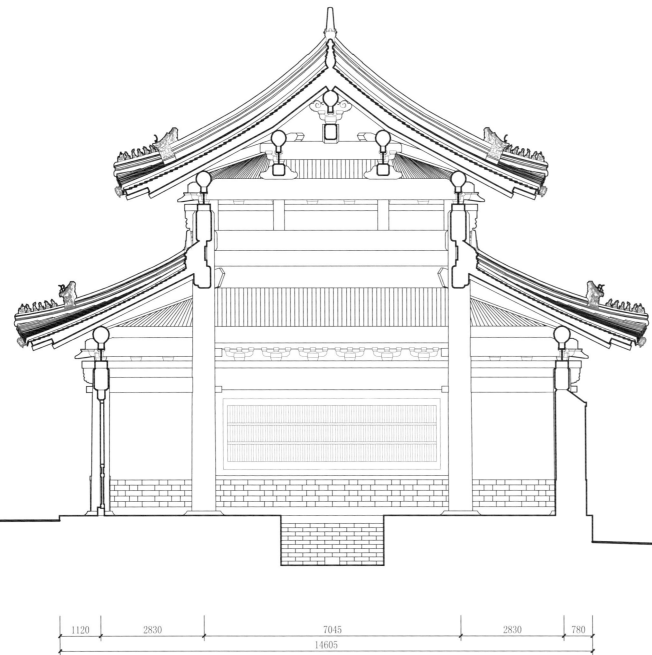

13.449

12.719

8.352
8.062

4.376
3.909

0.945

±0.000

-0.753

| 1120 | 2830 | 7045 | 2830 | 780 |

14605

北宰牲亭明间剖面图
Section of central-bay of North Pavilion of Immolation

0　　1　　　　　3m

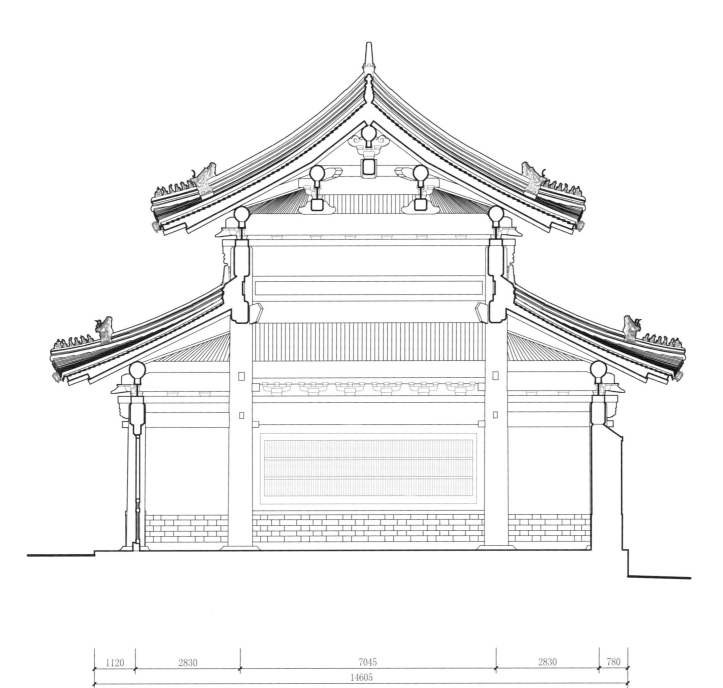

13.449

12.719

8.352
8.062

4.376
3.909

0.945

±0.000

-0.753

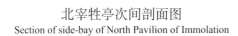

| 1120 | 2830 | 7045 | 2830 | 780 |

14605

北宰牲亭次间剖面图
Section of side-bay of North Pavilion of Immolation

0　1　　　3m

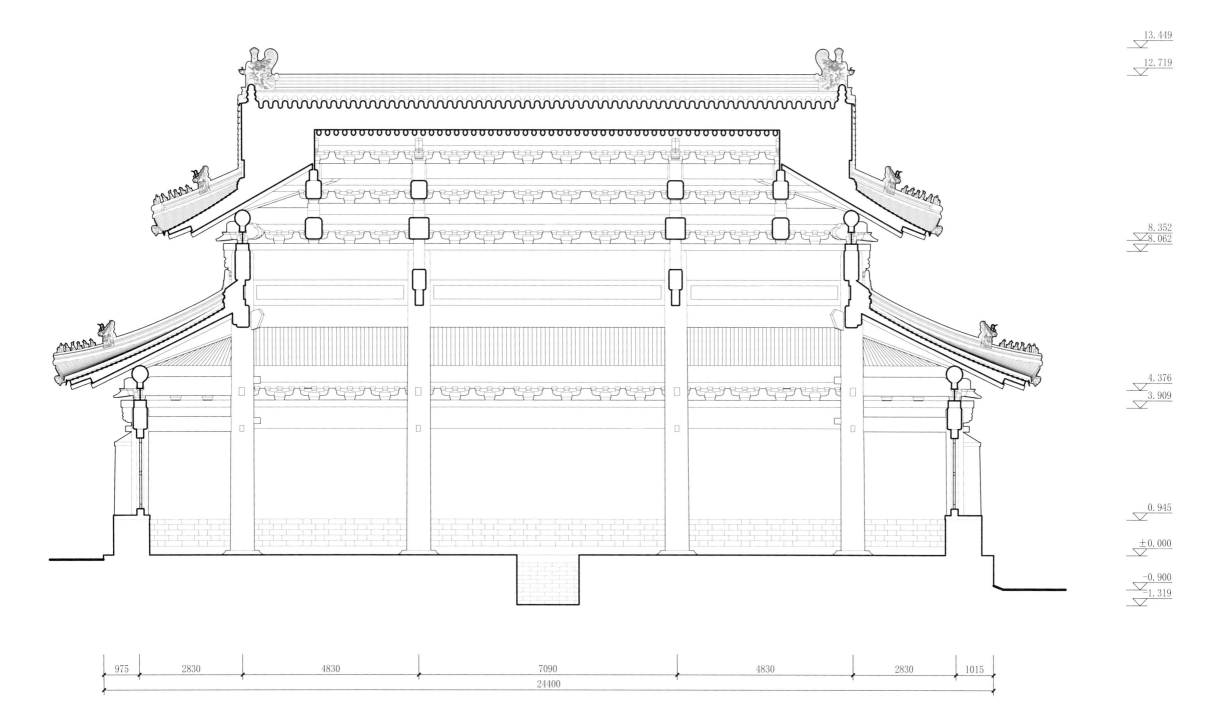

13.449
12.719

8.352
8.062

4.376
3.909

0.945

±0.000

-0.900
-1.319

975　2830　4830　7090　4830　2830　1015

24400

北宰牲亭纵剖面图
Longitudinal section of North Pavilion of Immolation

0　1　3m

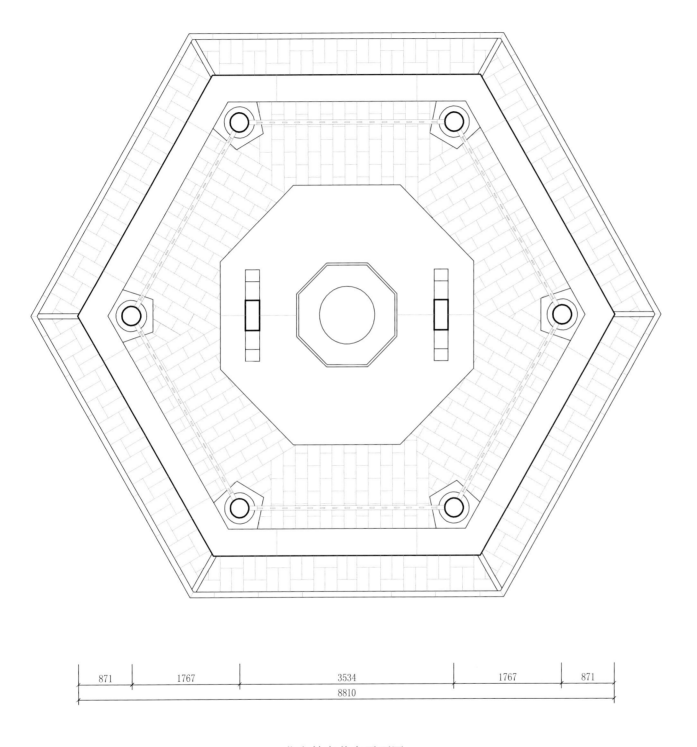

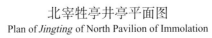

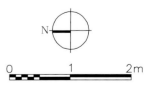

北宰牲亭井亭平面图
Plan of *Jingting* of North Pavilion of Immolation

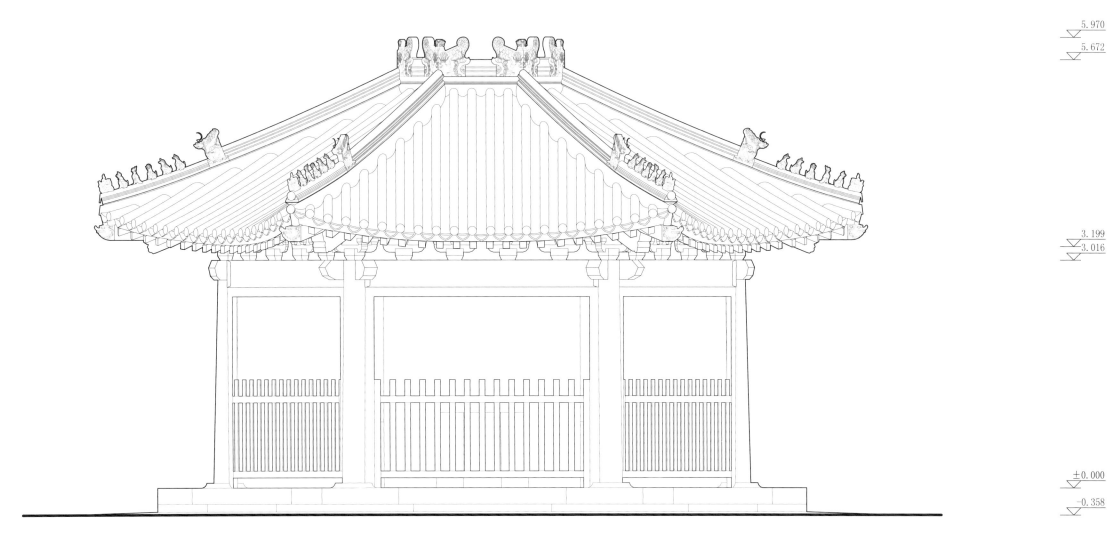

5.970
5.672

3.199
3.016

±0.000

-0.358

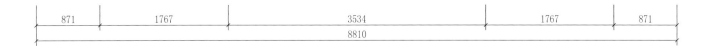

871 1767 3534 1767 871
8810

北宰牲亭井亭正立面图
Front elevation of *Jingting* of North Pavilion of Immolation

0 1 2m

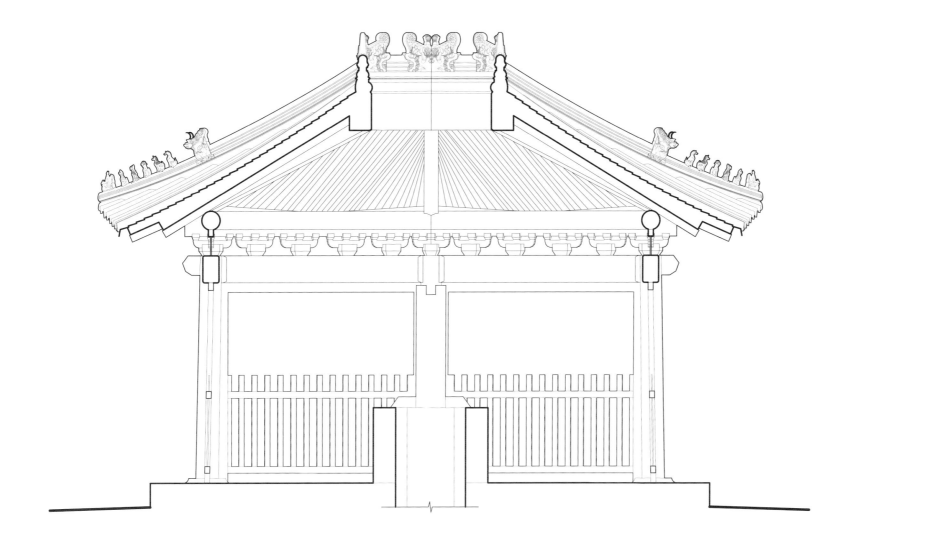
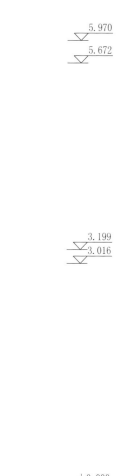

5.970
5.672

3.199
3.016

±0.000
−0.358

754 3061 3061 754
7630

北宰牲亭井亭横剖面图
Cross-section of *Jingting* of North Pavilion of Immolation

0 1 2m

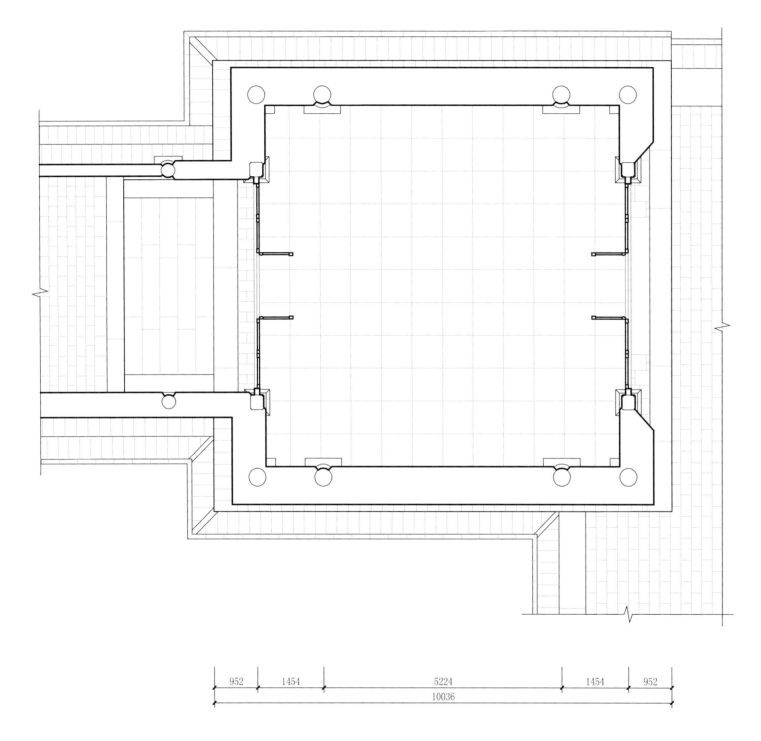

728
1583
4926
9518
1583
728

952　1454　5224　1454　952
10036

北宰牲亭门殿平面图
Plan of *Mendian* of North Pavilion of Immolation

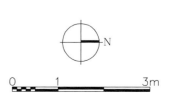

N

0　1　3m

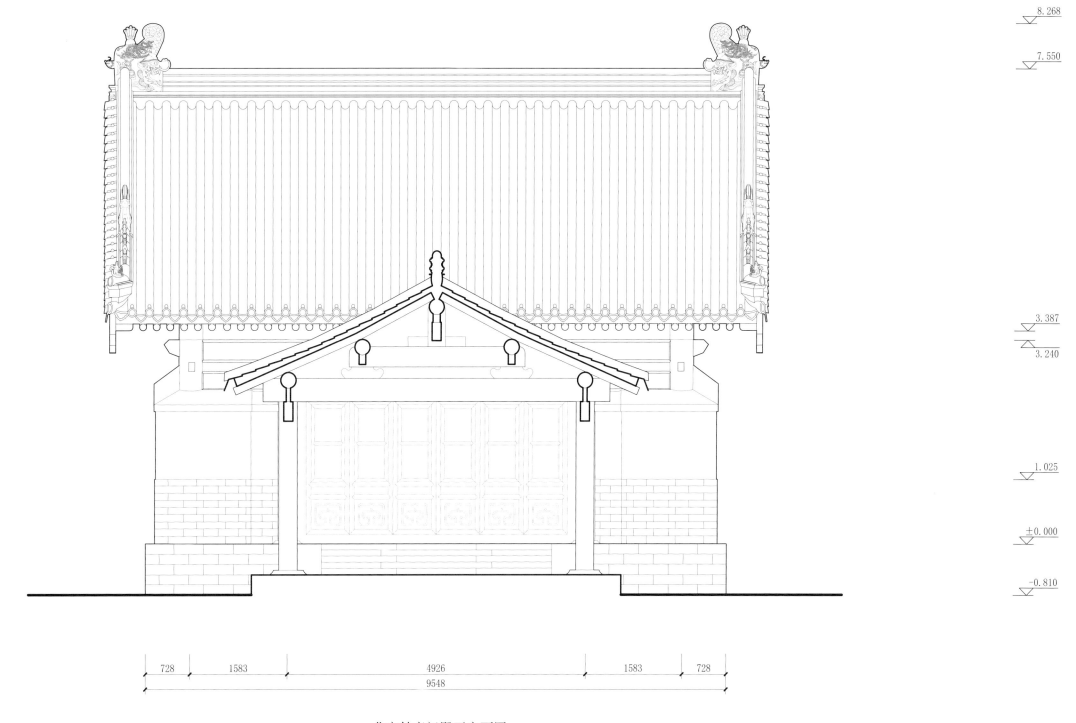

8. 268

7. 550

3. 387

3. 240

1. 025

±0. 000

−0. 810

| 728 | 1583 | 4926 | 1583 | 728 |

9548

北宰牲亭门殿正立面图
Front elevation of *Mendian* of North Pavilion of Immolation

0　　1　　2m

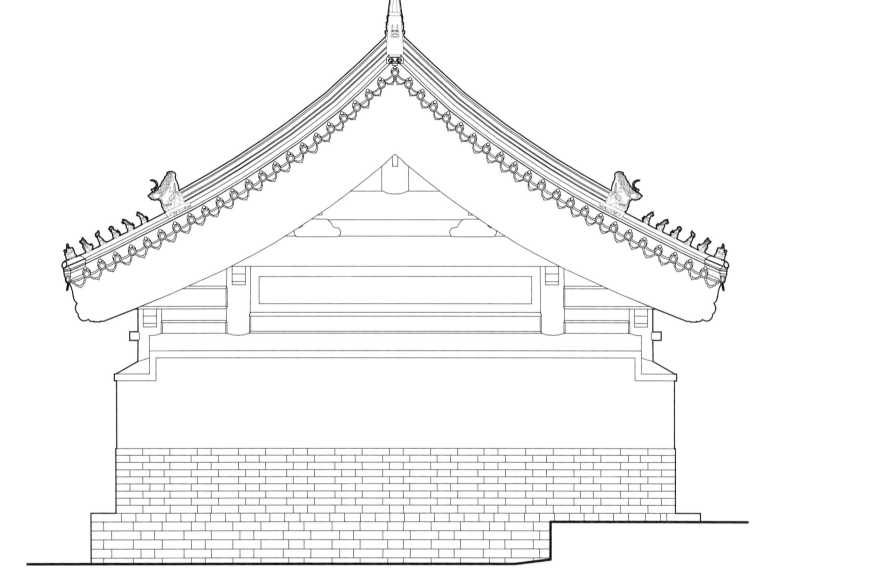

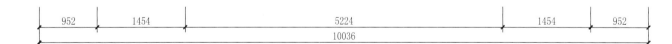

北宰牲亭门殿侧立面图
Side elevation of *Mendian* of North Pavilion of Immolation

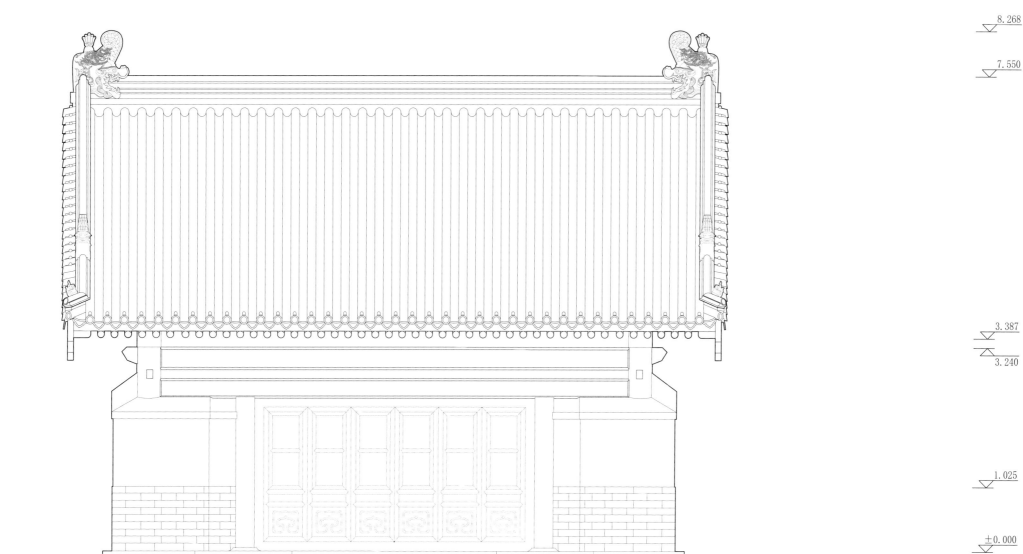

8.268

7.550

3.387

3.240

1.025

±0.000

−0.140

728   1583   4926   1583   728

9548

北宰牲亭门殿背立面图
Rear elevation of *Mendian* of North Pavilion of Immolation

0   1   2m

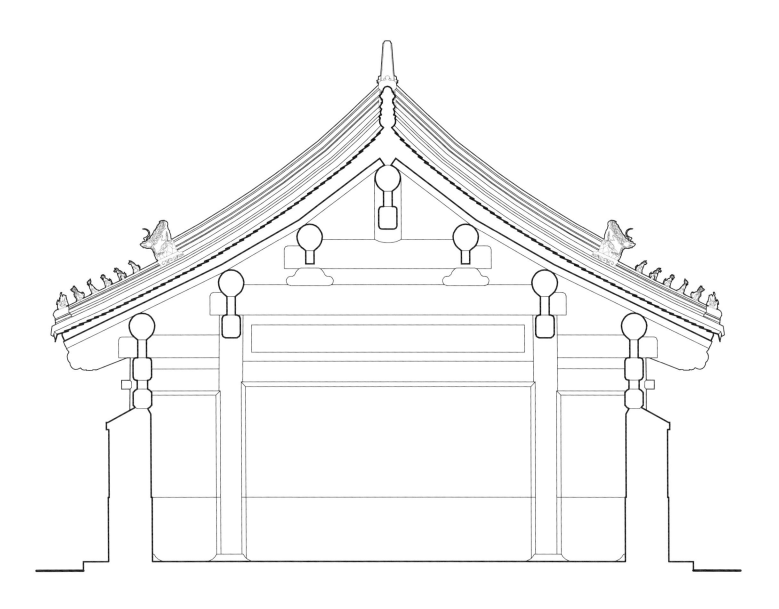

8.268

7.550

3.387

3.240

1.025

±0.000

-0.140

952　1454　5224　1454　952

10036

北宰牲亭门殿横剖面图

Cross-section of *Mendian* of North Pavilion of Immolation

0　1　2m

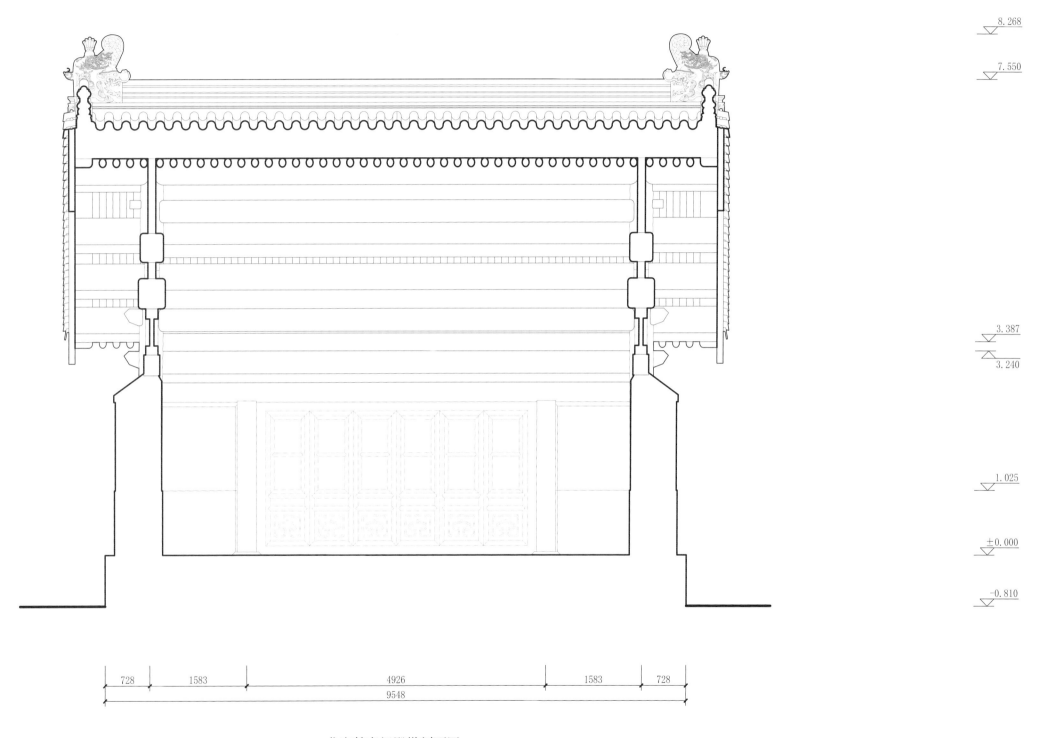

8. 268

7. 550

3. 387

3. 240

1. 025

±0. 000

-0. 810

728　1583　4926　1583　728

9548

北宰牲亭门殿纵剖面图
Longitudinal section of *Mendian* of North Pavilion of Immolation

0　1　2m

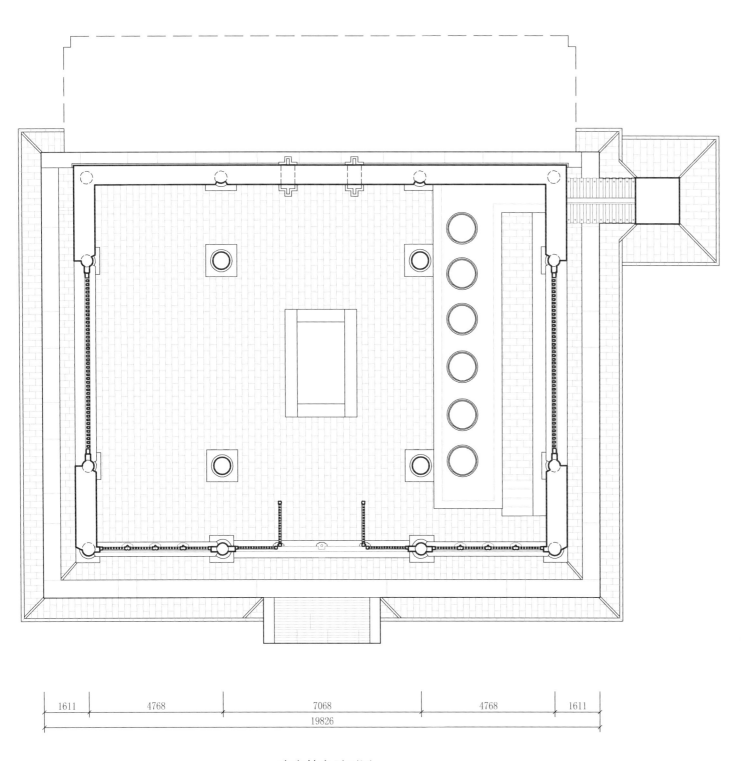

3997

873

2866

7061

15336

2866

1670

1611   4768   7068   4768   1611
19826

南宰牲亭平面图
Plan of South Pavilion of Immolation

N

0   1   3m

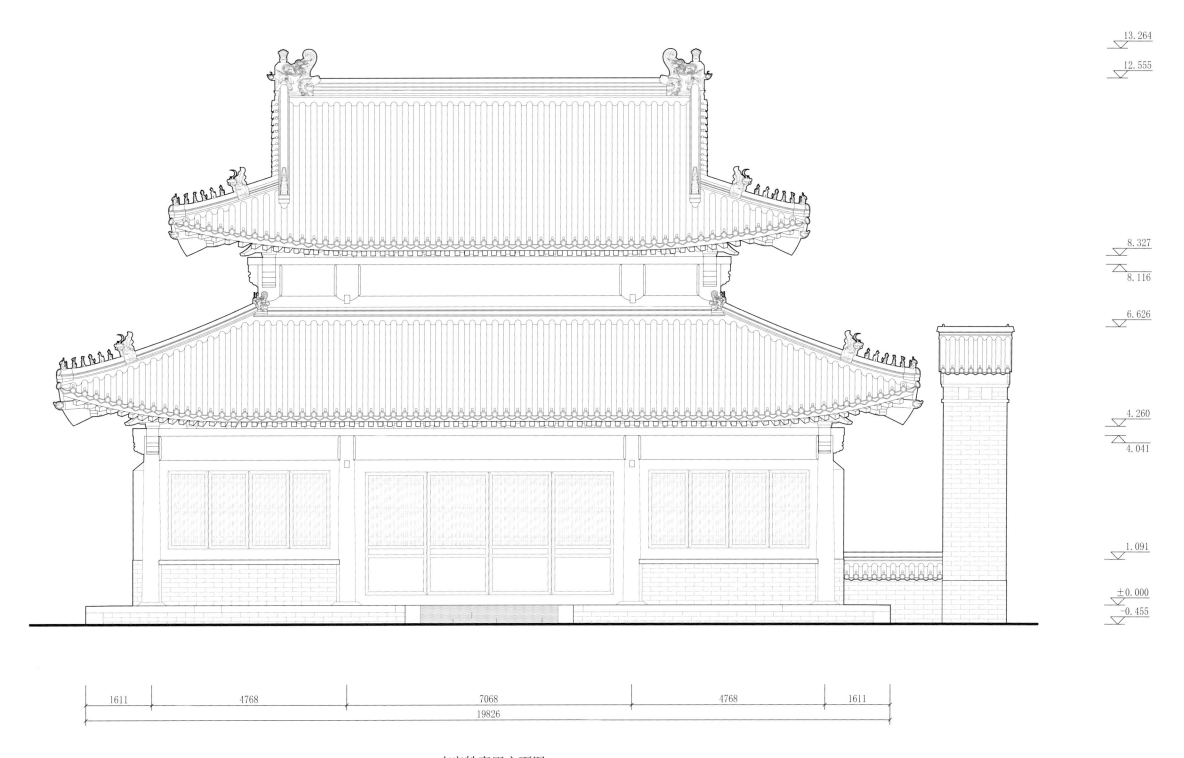

13.264

12.555

8.327

8.116

6.626

4.260

4.041

1.091

±0.000

-0.455

1611　　4768　　　　7068　　　　4768　　1611

19826

南宰牲亭正立面图
Front elevation of South Pavilion of Immolation

0　　1　　　　3m

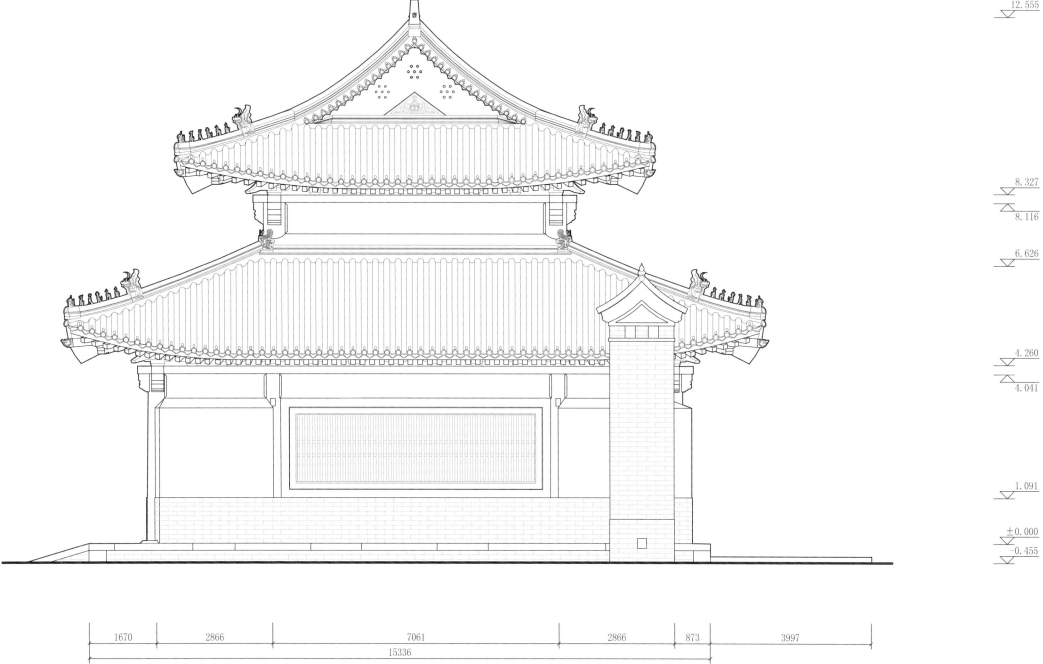

13.264

12.555

8.327

8.116

6.626

4.260

4.041

1.091

±0.000

-0.455

| 1670 | 2866 | 7061 | 2866 | 873 | 3997 |

15336

南宰牲亭侧立面图
Side elevation of South Pavilion of Immolation

0　　1　　　　3m

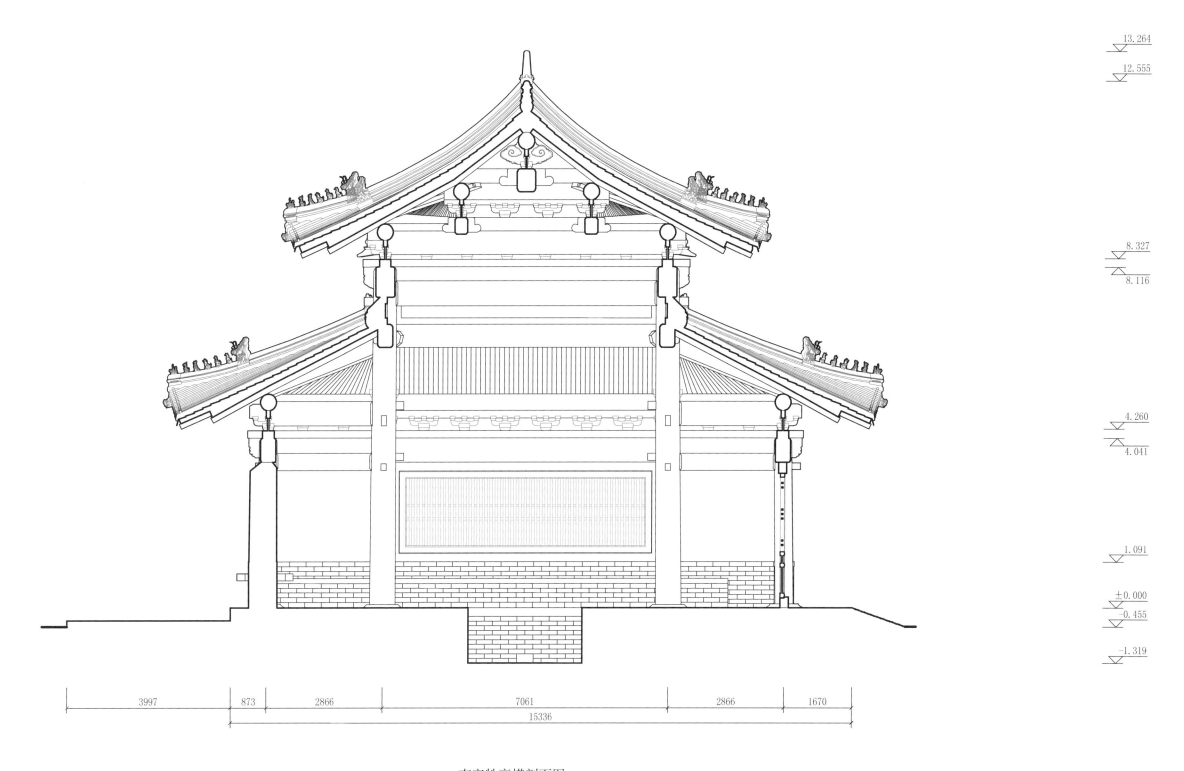

13.264

12.555

8.327

8.116

4.260

4.041

1.091

±0.000

-0.455

-1.319

| 3997 | 873 | 2866 | 7061 | 2866 | 1670 |

15336

南宰牲亭横剖面图
Cross-section of South Pavilion of Immolation

0 1 3m

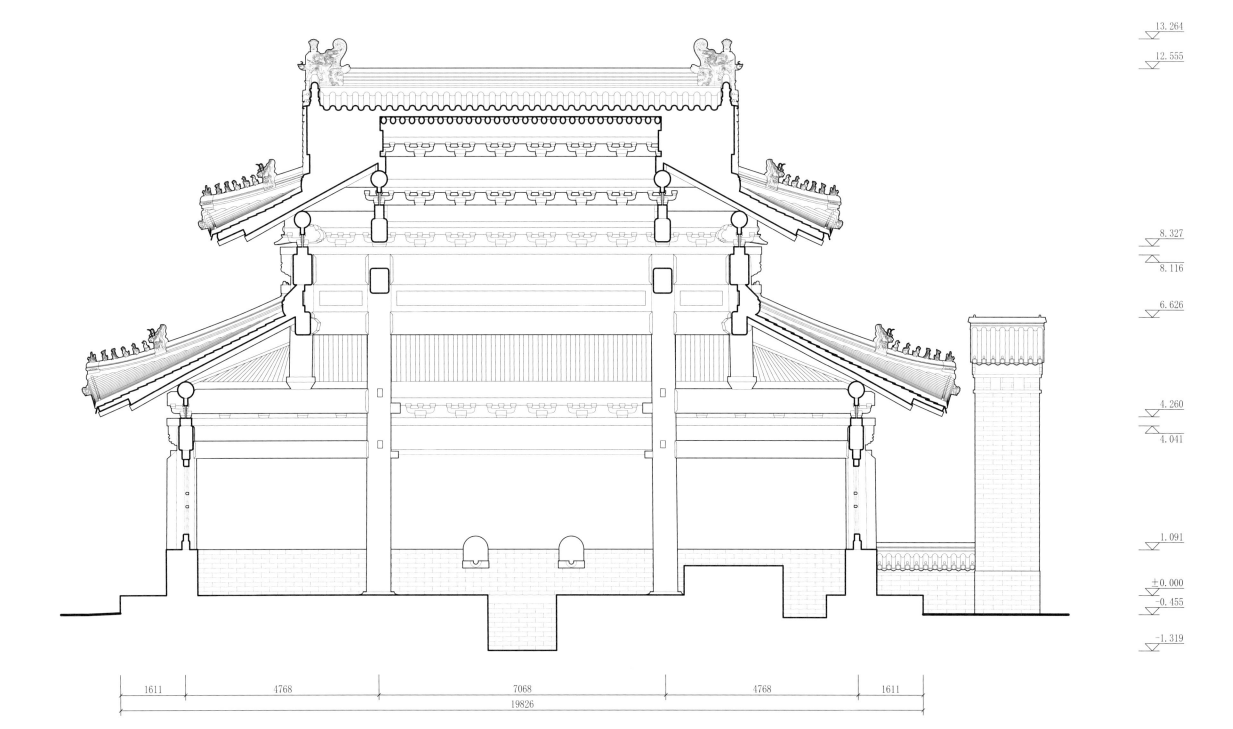

13.264

12.555

8.327

8.116

6.626

4.260

4.041

1.091

±0.000

-0.455

-1.319

1611　4768　7068　4768　1611

19826

南宰牲亭纵剖面图
Longitudinal section of South Pavilion of Immolation

0　1　3m

成贞门组群
Chengzhen Gate

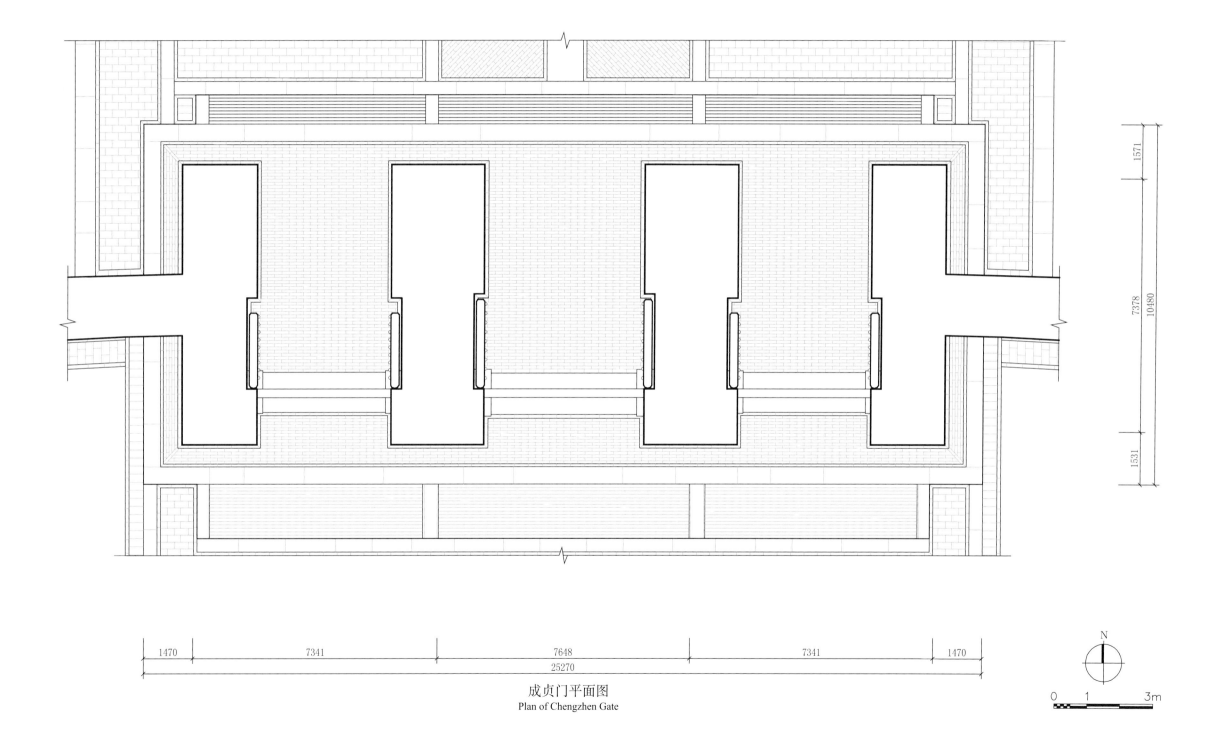

1571
7378
10480
1531

1470　7341　7648　7341　1470
25270

成贞门平面图
Plan of Chengzhen Gate

0　1　3m

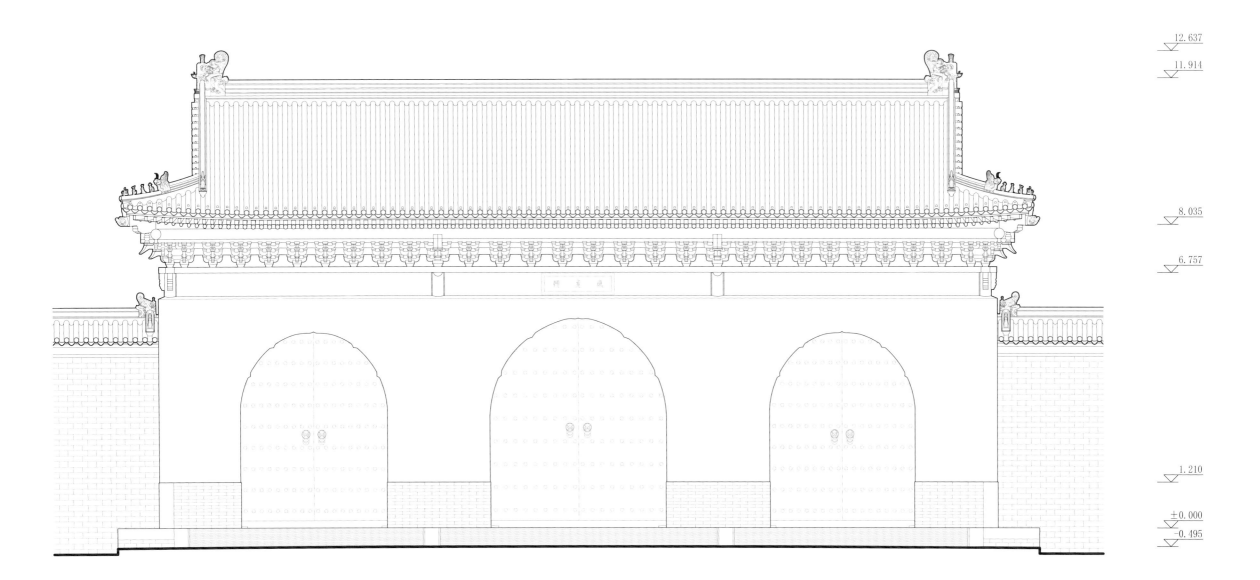

12.637

11.914

8.035

6.757

1.210

±0.000

−0.495

| 1470 | 7341 | 7648 | 7341 | 1470 |

25270

成贞门正立面图

Front elevation of Chengzhen Gate

0 1 3m

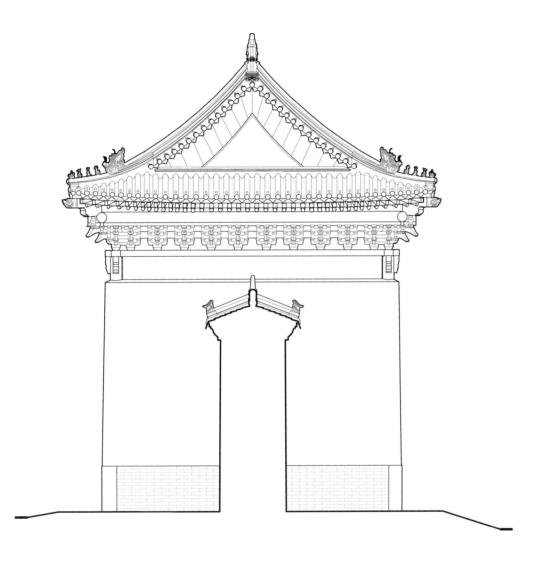

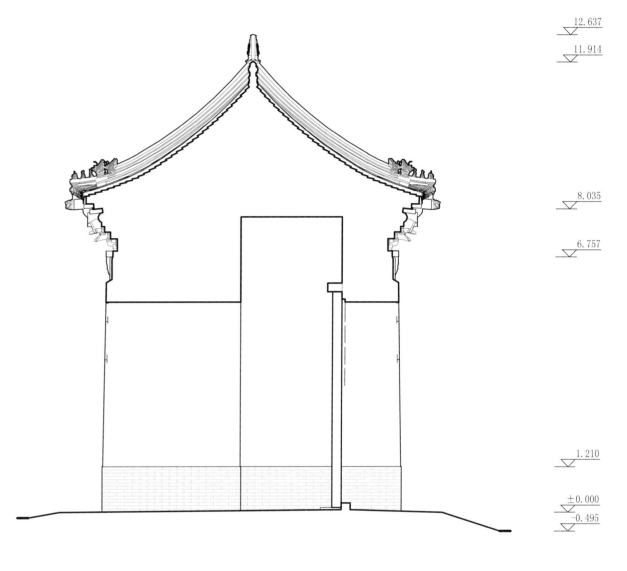

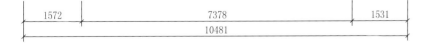

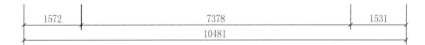

成贞门侧立面图
Side elevation of Chengzhen Gate

成贞门横剖面图
Section of Chengzhen Gate

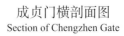

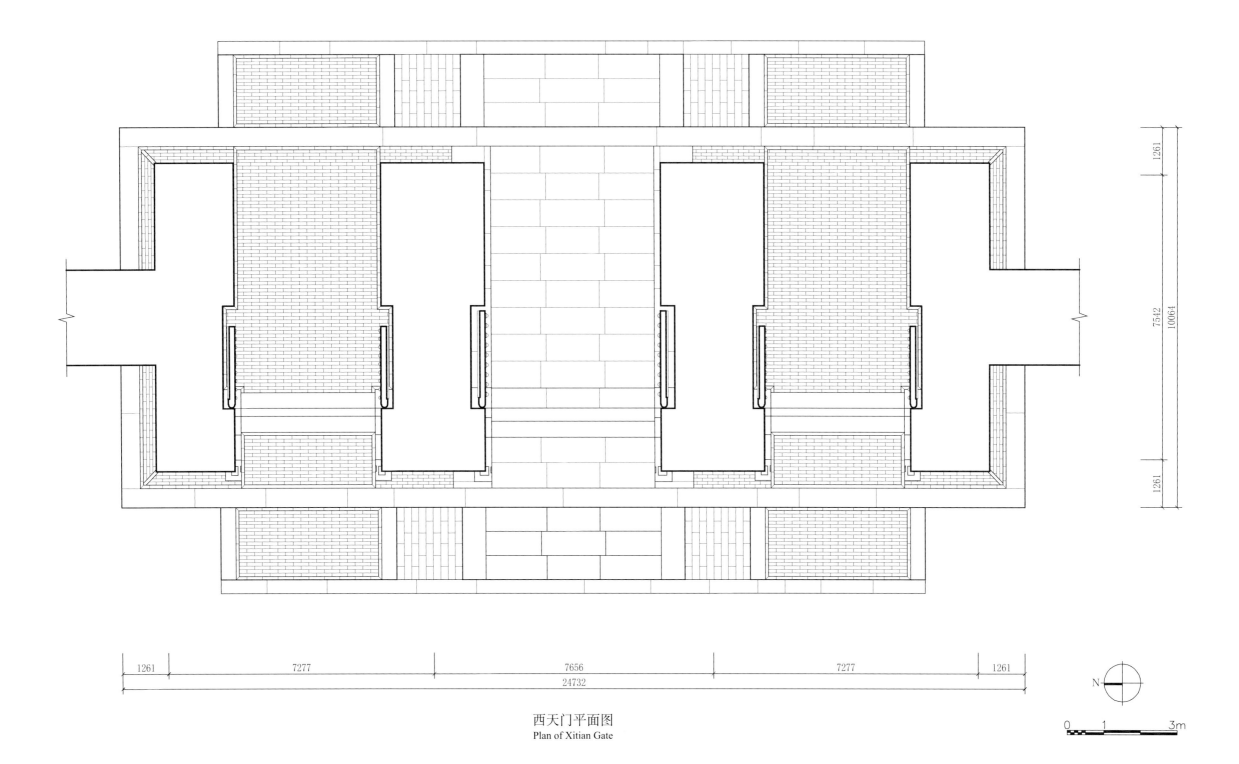

1261

1261

7542
10064

1261   7277   7656   7277   1261
24732

西天门平面图
Plan of Xitian Gate

N

0   1   3m

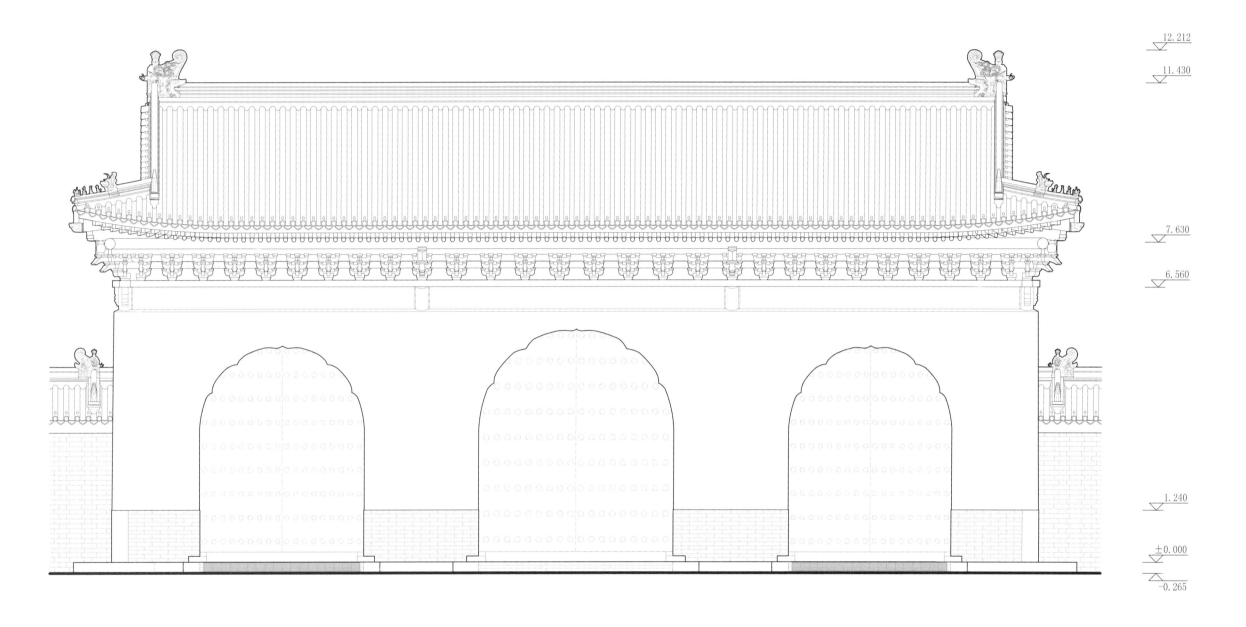

12.212

11.430

7.630

6.560

1.240

±0.000

-0.265

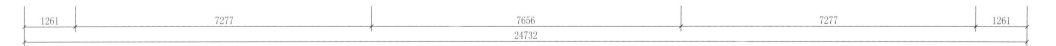

| 1261 | 7277 | 7656 | 7277 | 1261 |

24732

西天门正立面图
Front elevation of Xitian Gate

0 1 3m

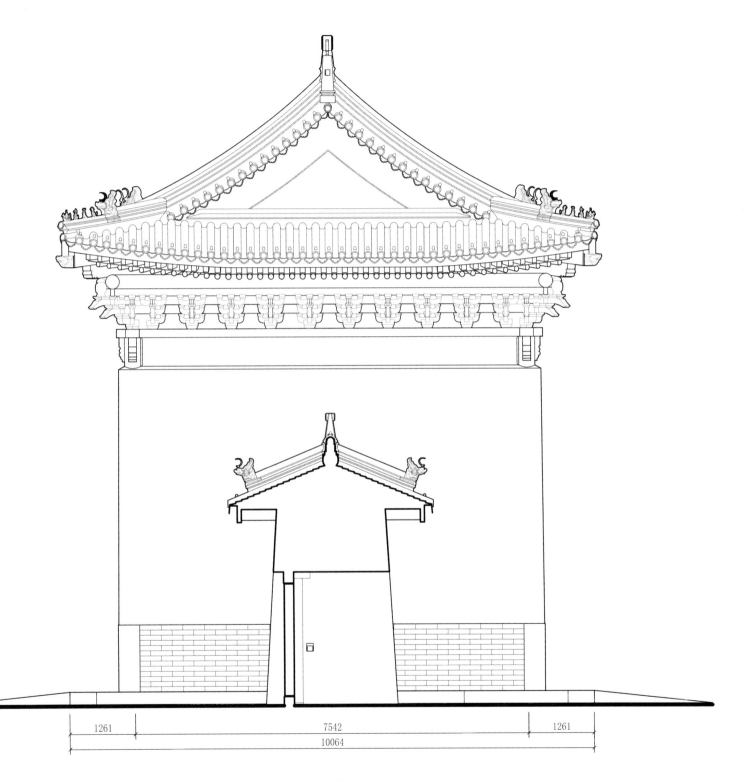

12.212

11.430

7.630

6.560

1.240

±0.000

-0.265

1261 7542 1261

10064

西天门侧立面图
Side elevation of Xitian Gate

0   1   2m

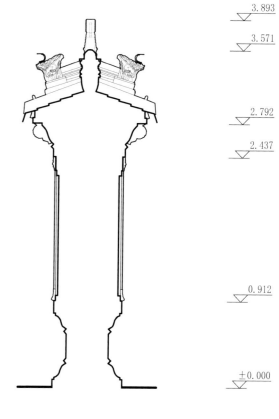

西天门影壁正立面图
Front elevation of *Yingbi* of Xitian Gate

西天门影壁侧立面图
Side elevation of *Yingbi* of Xitian Gate

西天门影壁横剖面图
Cross-section of *Yingbi* of Xitian Gate

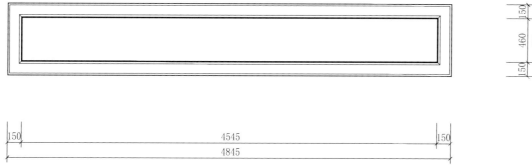

西天门影壁平面图
Plan of *Yingbi* of Xitian Gate

0   0.5   1m

参与天坛古建筑测绘及相关工作的人员名单

1986 年测绘

1984 级本科生：

韩向春　李少秋　程岩　张红红　张宏

1998 年测绘

教师：王其亨　张威　张玉坤　梁雪　洪再生　岳树信　白成军

博士研究生：吴葱

硕士研究生：庄岳　赵向东

1996 级本科生：

刘志杰　钟声　常征　陈跃东　侯俊峰　徐淼　吴涛　邱建伟　刘向晖

董建宁　秦昭　李凯　林林　李晨光　陈静　曲翠萃　赵玉红　王波

李修文　张一　马东明　梁航琳　周超　甄栋　刘迎坤　万树怡　刘向峰

郑小东　詹鹏　李涛　李强　奚凌晨　习皓　肖宇翔　许海燕　林志刚

胡威　陈君　石磊　徐鹏　柴镇硕　程涛　朱蕾　赵坚　周超

李笑　李淼　潘彤　沈苏　蔡良娃　胡劲松　薛忠燕　卞宇　杨明

徐世文　许熙巍　蒋丕彦　承昭　方轶　辜海玮　刘莹　唐栩　史丽川

阮步能　刘向明

1999 年测绘

教师：王其亨　张威　白雪海

硕士研究生：苏怡　姜东成　韩静　马航

1997 级本科生：

陈天泽　陈川　高悦　郭宁　潘旭东　董晶涛　郭勇宽　李兵　王焱

辛慧琴　梁建伟　熊潆　董岩　许铭（天津城建学院）

2005 年测绘

教师：王其亨　曹鹏　丁垚　白成军　吴葱

硕士研究生：李峥　吴晓冬　何蓓洁　陈芬芳　耿威　林佳

2001 级本科生：何蓉

2002 级本科生：刘瑜　于继成　高冉　谷方芳　唐涵

2003 级本科生：陈筱　王蕊佳

2006 年测绘

教师：王其亨　曹鹏　丁垚

博士研究生：张龙　吴琛

2003 级本科生：刘翔宇　石崧　马睿　徐萌

2015 年测绘

教师：曹鹏

硕士研究生：曹睿原　刘殿行　史展　孟晓静　周俊良

2011 级本科生：韩博　刘程明　王亚增　杨朝

2017 年测绘

教师：曹鹏

硕士研究生：谢怡明　杨莹　田甜　李东祖　江林燕　刘未达　周瀚

2015 级本科生：

朱翔宇　许琳　赵昕怡　郑佳茜　林郁　郑术闻　吴博萌　张欢

李璟　张镜荣　范家珲

测绘图整理出版参与人员

图审阅：王其亨　曹鹏

图纸整理修改：刘殿行　史展　曹睿原　孟晓静　周俊良　杨莹　田甜

谢怡明　李东祖　江林燕　冯亚欣　刘欣佳　王艺璇　盛举艳　满兵兵　姚禹舍

刘程明　韩博　王亚增　杨朝　张涛

北京市天坛公园管理处：李高　于辉　林冬生　王恩铭　李璐　申博

刘上华　刘垚　刘毅

翻译人员

英文翻译：郭涵　周彦邦

英文审校：刘仁皓

List of Participants Involved in Surveying and Related Works

Survey Time: 1986

Undergraduate Students (Class 1984): HAN Xiangchun, LI Shaoqiu, CHENG Yan, ZHANG Honghong, ZHANG Hong

Survey Time: 1998

Supervisors: WANG Qiheng, ZHANG Wei, ZHANG Yukun, LIANG Xue, HONG Zaisheng, YUE Shuxin, BAI Chengjun

Doctoral Student: WU Cong

Master Students: ZHUANG Yue, ZHAO Xiangdong

Undergraduate Students (Class 1996): LIU Zhijie, ZHONG Sheng, CHANG Zheng, CHEN Yuedong, HOU Junfeng, XU Miao, WU Tao, YUE Jianwei, LIU Xianghui, DONG Jianning, QIN Zhao, LI Kai, LIN Lin, LI Chenguang, CHEN Jing, QU Cuicui, ZHAO Yuhong, WANG Bo, LI Xiuwen, ZHANG Yi, MA Dongming, LIANG Hanglin, ZHOU Chao, ZHEN Dong, LIU Yingkun, WAN Shuyi, LIU Xiangfeng, ZHENG Xiaodong, ZHAN Peng, LI Tao, LI Qiang, XI Lingchen, XI Hao, XIAO Yuxiang, XU Haiyan, LIN Zhigang, HU Wei, CHEN Jun, SHI Lei, XU Peng, CHAI Zhenshuo, CHEN Tao, ZHU Lei, ZHAO Jian, ZHOU Chao, LI Xiao, LI Miao, PAN Tong, SHEN Su, CAI Liangwa, HU Jinsong, XUE Zhongyan, BIAN Yu, YANG Ming, XU Shiwen, XU Xiwei, JIANG Peiyan, CHENG Zhao, FANG Yi, GU Haiwei, LIU Ying, TANG Xu, SHI Lichuan, RUAN Buneng, LIU Xiangming

Survey Time: 1999

Supervisors: WANG Qiheng, ZHANG Wei, BAI Xuehai

Master Students: SU Yi, JIANG Dongcheng, HAN Jing, MA Hang

Undergraduate Students (Class 1997): CHEN Tianze, CHEN Chuan, GAO Yue, GUO Ning, PAN Xudong, DONG Jingtao, GUO Yongkuan, LI Bing, WANG Yan, XIN Huiqin, LIANG Jianwei, XIONG Ying, DONG Yan, Xu Ming (Tianjin Chengjian University)

Survey Time: 2005

Supervisors: WANG Qiheng, CAO Peng, DING Yao, BAI Chengjun, WU Cong

Master Students: LI Zheng, WU Xiaodong, HE Beijie, CHEN Fenfang, GENG Wei, LIN Jia

Undergraduate Student (Class 2001): HE Rong

Undergraduate Students (Class 2002): LIU Yu, YU Jicheng, GAO Ran, GU Fangfang, TANG Han

Undergraduate Students(Class 2003): CHEN Xiao, WANG Ruijia

Survey Time: 2006

Supervisors: WANG Qiheng, CAO Peng, DING Yao

Doctoral Students: ZHANG Long, WU Chen

Undergraduate Students (Class 2003): LIU Xiangyu, SHI Song, MA Rui, XU Meng

Survey Time: 2015

Supervisor: CAO Peng

Master Students: CAO Ruiyuan, LIU Dianhang, SHI Zhan, MENG Xiaojing, ZHOU Junliang

Undergraduate Students (Class 2011): HAN Bo, LIU Chengming, WANG Yazeng, YANG Zhao

Survey Time: 2017

Supervisors: CAO Peng

Master Students: XIE Yiming, YANG Ying, TIAN Tian, LI Dongzu, JIANG Linyan, LIU Weida, ZHOU Han

Undergraduate Students (Class 2015): ZHU Xiangyu, XU Lin, ZHAO Xinyi, ZHENG Jiaqian, LIN Yu, ZHENG Shuwen, WU Bomeng, ZHANG Huan, LI Jing, ZHANG Jingrong, FAN Jiahui

Editor of Drawings

Supervisors：WANG Qiheng，CAO Peng

Revision and Arrangement Personnel：LIU Dianxing, SHI Zhan, CAO Ruiyuan，MENG Xiaojing,，ZHOU Junliang，YANG Ying，TIAN Tian, XIE Yiming，LI Dongzu，JIANG Linyan，FENG Yaxin，LIU Xinjia，WANG Yixuan，SHENG Juyan，MAN Bingbing，YAO Yuhan，LIU Chengming，HAN Bo，WANG Yazeng，YANG Zhao，ZHANG Tao

The Administration of the Temple of Heaven：LI Gao，YU Hui，LIN Dongsheng，WANG Enming，LI Lu，SHEN Bo，LIU Shanghua，LIU Yao，LIU Yi

Editor of Translation

Translator：GUO Han，CHOU Yen-Pang

Proofreader：LIU Renhao

图书在版编目（CIP）数据

天坛＝The Temple of Heaven：汉英对照 / 王其
亨主编；曹鹏编著；天津大学建筑学院，北京市天坛公
园管理处编写.—北京：中国建筑工业出版社，2019.12
（中国古建筑测绘大系.坛庙建筑）
ISBN 978-7-112-24560-4

Ⅰ.①天… Ⅱ.①王… ②曹… ③天… ④北… Ⅲ.
①天坛—建筑艺术—图集 Ⅳ.① TU-092.2

中国版本图书馆CIP数据核字（2019）第286226号

丛书策划 / 王莉慧
责任编辑 / 李　鸽　李　婧
书籍设计 / 付金红
责任校对 / 王　烨

中国古建筑测绘大系·坛庙建筑

天坛

天津大学建筑学院
北京市天坛公园管理处　编写

王其亨　主编　曹鹏　编著

Traditional Chinese Architecture Surveying and Mapping Series: Temples Architecture
The Temple of Heaven
Compiled by School of Architecture, Tianjin University & The Administration of The Temple of Heaven
Chief Edited by WANG Qiheng, Edited by CAO Peng

*

中国建筑工业出版社出版、发行（北京海淀三里河路 9 号）
各地新华书店、建筑书店经销
北京方舟正佳图文设计有限公司制版
北京雅昌艺术印刷有限公司印刷

*

开本：787 毫米 ×1092 毫米　横 1/8　印张：32½　字数：596 千字
2021 年 9 月第一版　2021 年 9 月第一次印刷
定价：**258.00** 元
ISBN 978-7-112-24560-4
（35214）